THOMPSON SUBMACHINE GUNS

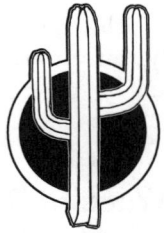

Desert Publications
El Dorado, AR 71731-1751 U. S. A.

Thompson Submachine Guns

© 1978 by Desert Publications
215 S. Washington Ave.
El Dorado, AR 71730 U. S. A.
1-800-852-4445
info@deltapress.com
Printed in U. S. A

ISBN 0-87947-031-3
12 11 10 9 8 7 6

Desert Publication is a division of
The DELTA GROUP, Ltd.
Direct all inquiries & orders to the above address.

CONTENTS

PART ONE

The

THOMPSON SUBMACHINE GUN

Model 1928 ·45 Calibre

Fully illustrated with 17 Photographic Plates

ALDERSHOT
GALE & POLDEN LTD

PUBLISHERS' FOREWORD

This book is not meant to replace the official publications on the gun, but rather to be read in conjunction with them

The publishers have received so many inquiries for a comprehensive booklet on the Thompson Submachine Gun that this booklet has been written and fully illustrated in an attempt to give as much information in as condensed form as possible.

By studying this booklet with the gun by his side, so that the actual parts referred to and illustrated in its pages can be readily identified, the reader will find that he will be able to understand the operation of the mechanism and the nomenclature of parts, and to strip and handle the gun in a very short time.

Instructors should note that this booklet is intended to meet the requirements of those who need a comprehensive handbook on the gun. It is not intended to take the place of the official publication which lays down the lessons and drill to be taught to those who will be armed with the weapon. If this booklet is used as a reference for this purpose the publishers draw attention to those sections and paragraphs marked with a † which give detail information that is not essential to include in lessons, save in those cases where training time permits or it is felt their inclusion necessary. The English nomenclature of parts (*i.e.*, those in italics in the captions to illustrations) only need be taught.

Gale & Polden Ltd.
LONDON & ALDERSHOT

CONTENTS

ILLUSTRATIONS

English names of parts are shown in italics.

American names of parts in Roman type.

FIRING FROM THE SHOULDER

Note position of feet, body well forward, shoulder into gun,
right elbow well up, right leg braced.

THE THOMPSON SUBMACHINE GUN
Mechanism Made Easy

DATA

GUN

Overall length of gun with stock	$33\frac{3}{4}$ ins.
Overall length of gun without stock ...	$25\frac{1}{4}$ ins.
Weight (without magazine)	10 lbs.
Weight of gun with drum magazine, loaded 50 rounds	$14\frac{3}{4}$ lbs.
Weight of gun with box magazine, loaded 20 rounds	$11\frac{1}{4}$ lbs.
Barrel length without Cutts compensator	$10\frac{1}{2}$ ins.
Barrel length with Cutts compensator ...	$12\frac{1}{2}$ ins.
Rifling right-hand one turn in 16 ins. ...	
Rate of fire, semi-automatic (single-aimed shots) per minute	Up to 100 shots.
Cyclic rate of fire, fully automatic (approx.)	700 rds. per min.

MAGAZINES

Drum magazine capacity	50 rds.
Drum magazine weight, fully loaded ...	$4\frac{3}{4}$ lbs.
Box magazine capacity	20 rds.
Box magazine weight, fully loaded ...	$1\frac{1}{4}$ lbs.

AMMUNITION

.45 calibre rimless auto pistol cartridge.

Weight of bullet	230 grains.
Weight of cartridge	324 grains.
Weight of 50 cartridges	$2\frac{1}{4}$ lbs.

SIGHTS

Foresight blade.
Backsight adjustable slide on leaf graduated 50-600 yards.
Open battle sight (aperture in position with leaf lowered) sighted for 50 yards.

GENERAL DESCRIPTION

The Thompson Submachine Gun is a small handy automatic weapon capable of delivering a high rate of fire.

It is provided with a change lever on the trigger mechanism whereby either fully automatic fire (continuous bursts of fire) or semi-automatic fire (single shots) may be delivered.

Fins are turned on the barrel which assist to dissipate the heat generated when the gun is fired. The barrel, however, soon becomes too hot to touch and a special forend grip is fitted under the barrel so that the hand holding same is kept well away from the barrel.

PRINCIPLE OF OPERATION

The gun is recoil operated by the backward thrust of the cartridge case which occurs when the charge explodes. There is no *positive* locking action of the bolt ; the delay in the opening of the breech, so that the cartridge is given the necessary support in the chamber at the moment the gun is fired, works on the principle known as the Hesitation or Delayed Action System.

†MECHANISM

On the firing of the gun the expanding gases of the exploded charge exert pressure—

(A) against the base of the bullet to drive it through the bore of the barrel ;

(B) against the walls of the cartridge case which obturate (are expanded) against the walls of the chamber ;

(C) against the base of the cartridge case.

REARWARD MOVEMENT OF THE BOLT

The pressure against the base of the cartridge case overcomes the mechanism on the gun which provides the principal element of friction to delay the opening of the breech. When the breech is opened the spent case is projected rearward from the chamber, driving the bolt with it to the rear, until the base of the spent case strikes against the ejector and is ejected through the ejection port in the body, clear of the gun.

The bolt in its movement to the rear compresses the recoil spring.

6

If the change lever is in " Single Shot " position (or in the " Full Auto " position, and pressure on trigger is released) the sear of the trigger mechanism engages the bent in the bolt, thereby the bolt is held back and is prevented from being driven forward under the pressure of the compressed recoil spring.

FORWARD MOVEMENT OF THE BOLT

On the trigger being pressed the sear disengages the bent of the bolt, the compressed recoil spring reasserts itself, driving the bolt forward. The face of the bolt strikes the base of the next cartridge in the magazine (which has been fed into position by the magazine spring), carrying it into the chamber. The extractor springs into the grooved base of the cartridge.

As the bolt reaches its fully forward position home—*i.e.*, the breech is closed—the bottom forward end of the triangular hammer strikes against an abutment inside the front end of the receiver. The hammer being pivoted on its pin at its centre, the top forward end of the hammer strikes a forward blow to the rear end of the firing pin, with which it is in contact. The firing pin is thus driven forward and the striker hits the cap of the cartridge in the chamber and fires the round.

As the bolt approaches its foremost position the lugs on the H-piece engage their grooves in the body which, together with the pressure of the recoil spring imparted to the H-piece through the medium of the slot in the cocking handle (which is engaged with the bridge of the H-piece), drives the H-piece downwards into its original position again.

COCKING HANDLE.

The cocking handle rests in the bolt and is free to slide therein ; the inclined slot between its two fingers engages the H-piece bridge (or bar). The front end of the recoil spring is housed in a hole inside the cocking handle.

BREECH OILER

The breech oiler consists of a spring which holds oil-saturated felt pads. In its position in the body, the oil-saturated felt pads lubricate the lugs of the H-piece and help to keep the sides of the bolt lubricated as the bolt reciprocates backwards and forwards.

DELAYING ACTION MECHANISM

On explosion of the charge the resultant backward force that the spent case exerts against the face of the bolt is transmitted to the

7

H-piece which is situated in its 70° inclined slot in the bolt and so to the lugs of the H-piece which are engaged in short 45° recessed grooves in the body.

Thus resistance to the bolt's backward movement is encountered, because the lugs of the H-piece in engagement with the short 45° inclined slot in the body lift the H-piece ; resistance to this lifting action occurs because the front inclined face of the H-piece meets the rear face of the 70° inclined slot in the bolt. The rising of the H-piece is further delayed because its bridge engages the slot in the cocking handle which is set to the rear at an angle of 10° from the vertical.

The direction of the movement of the H-piece as a result of the movement of these components is upwards and backwards.

The above described delaying action prevents the bolt from moving to the rear during the period of high chamber pressure.

The adhesion of the frictional surfaces thus maintains the H-piece in a fixed position until the pressure in the chamber has so far dropped to the point when there is just enough residual pressure to impart the necessary momentum to drive the bolt to the rear and fully compress the recoil spring.

Plate II opposite shows the inclined surfaces of the slot in the cocking handle, the H-piece, the inclined slot in bolt and the angle of the inclined recessed grooves in the body.

CUTTS COMPENSATOR

The Cutts compensator (2, Plate III, page 11) is designed so that it compensates a tendency for the barrel to rise upwards, and also it reduces the amount of free recoil of the gun. This compensating action takes place as follows :—

The gases which are at high pressure behind the bullet immediately on the latter leaving the muzzle of the barrel, escape into the outer air through the orifices at the top of the compensator. The result is the blast of gases finding their way out gives a compensating thrust downwards on the compensator and so keeps the muzzle of the gun down.

At the same time the gases tend to blow the compensator off the muzzle of the gun ; the actual effect is to give a forward thrust to the gun, thus giving compensating action to the backward recoil of the gun.

NOTE.—Firing tends to fill the orifices of the compensator with residue from the burning powder charge ; it is important, in order that the full effect of the compensator is obtained, that the orifices should always be kept clean.

8

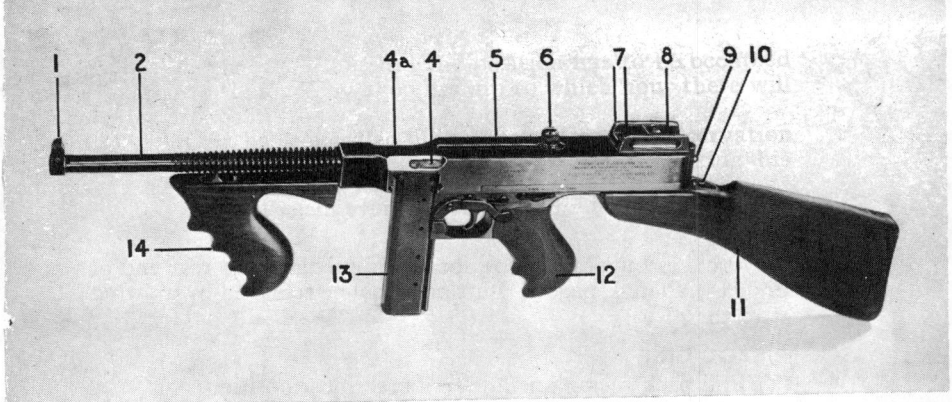

Plate I

1 *Foresight.*	8 *Backsight.*
2 *Barrel.*	9 Projection of buffer rod.
4 Ejector.	10 *Butt catch.*
4a Ejection port.	11 *Butt.*
5 *Body.*	12 *Pistol Grip.*
6 *Cocking handle.*	13 Box magazine (20 rounds
7 Backsight protector ramps.	capacity).
	14 Fore grip.

Plate II

Showing the angles of the inclined surfaces referred to in text.

5 *Body.*

5a Inclined recessed groove in *Body* for 22a.

6 *Cocking handle.*

16 Actuator.

16a Slot in actuator for bridge of H-piece.

22 H-*piece.*

22a Lugs of H-piece.

17 *Bolt.*

17a Inclined slot in bolt for H-piece.

20 Hammer.

21 Hammer pin.

23a Groove in bolt for extractor.

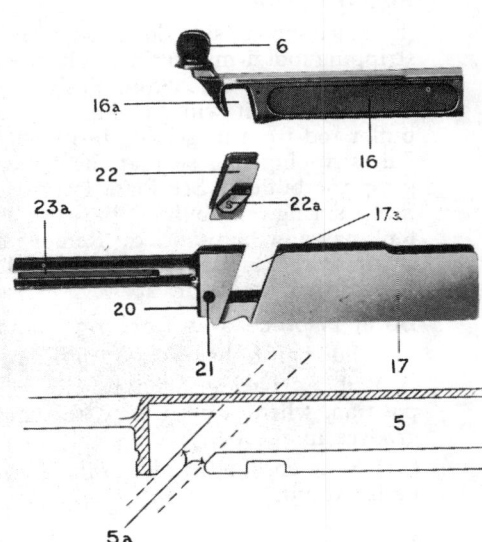

STRIPPING

(1) Ensure gun is unloaded and remove magazine.

BUTT

(2) Press in butt catch (10), located immediately to rear end of body (5), Plate I, page 9. Butt may now be removed by drawing it to the rear.

PISTOL GRIP

(3) Pull back cocking handle to its rearmost position.

(4) Put safety catch to " Fire " position.

(5) Set change lever to " Full Auto."

(6) Return bolt to its fully forward closed position, as follows : With the left hand gripping cocking handle in its rear position, press the trigger with the index finger of the right hand. The left hand allows the cocking handle to return slowly forward until the bolt is right home.

(7) Place gun upside down on a table or knees, press in stud (15, Plate III, opposite) and with it depressed, tap frame rearward a short distance. Now grasp the body in one hand and the pistol grip in the other and pull the trigger, and with it held depressed, the pistol grip can be slid off the body to the rear.

THE BODY GROUP

Recoil Spring

(8) Place gun upside down on a table, grasp the recoil spring stripping tool firmly in the left hand and insert it into its hole in the front end of buffer rod. Push the stripping tool towards the bolt as far as it will go. This will withdraw the rear end of buffer rod from its seating in its hole at the rear end of body. Tilt stripping tool so that the fingers of the right hand may grasp the buffer. (See Plate IV, page 13, illustration B.) The recoil spring, the buffer fibre disc and rod and the stripping tool may now be removed. Remove buffer fibre disc. Remove stripping tool, and whilst doing this hold buffer rod and spring so that they do not fly apart.

Bolt, H-piece and Cocking Handle.

(9) Slide bolt to its rearmost position, when it can be lifted out.

(10) Slide cocking handle with H-piece to their foremost position, when H-piece may be removed through its inclined grooves in the body.

(11) Slide cocking handle to its rearmost position, when it can be lifted out.

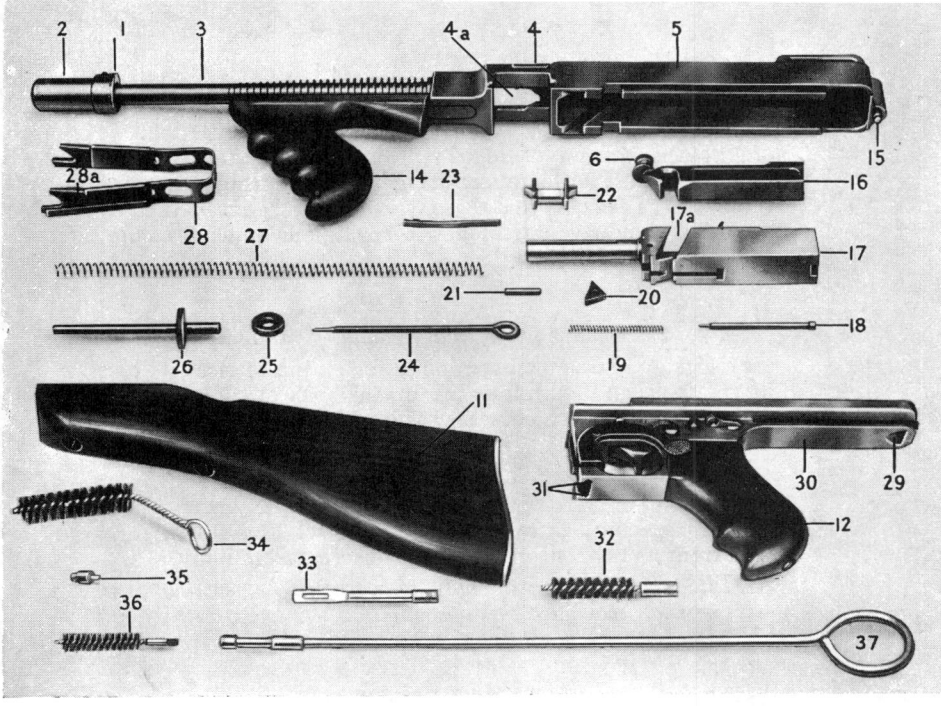

Plate III

★ *Parts marked thus if not issued are replaced by pull-through and gauze.*
(*See Note,* p. 24.)

Firing Pin

(12) With the end of recoil spring stripping tool push out hammer pin. The hammer, firing pin and spring can now be removed. (Take care that these parts do not fly out on their own accord as they have a tendency to spring out under pressure of the firing pin spring.)

†Extractor

(need not be removed for ordinary cleaning purposes)

(13) Lift up extractor just enough for its lug to clear its recess in the bolt and withdraw it forward. Take care not to lift extractor higher than necessary as it may break or become bent.

†Ejector

(need not be removed for ordinary cleaning purposes)

(14) Lift ejector leaf from its recess in body. This will disengage stud from its depression in body ; unscrew from body.

NOTE.—The ejector leaf must not be lifted higher than is just necessary to accomplish this operation.

†Breech Oiler

(need not be removed for ordinary cleaning purposes)

(15) Press the fingers of the breech oiler together sufficiently to clear the undercut recess in receiver, when it can be removed.

STRIPPED PARTS

The parts of the gun now stripped as described above 1-12 are illustrated in Plate III, page 11, and comprise all the parts necessary to be stripped for ordinary cleaning purposes, with the exception of the extractor, ejector and breech oiler, which do not always need to be removed for this.

†STRIPPING OF THE PISTOL GRIP GROUP

As it is not necessary to strip the pistol grip for ordinary cleaning purposes, no description is given of the operations necessary to accomplish this.

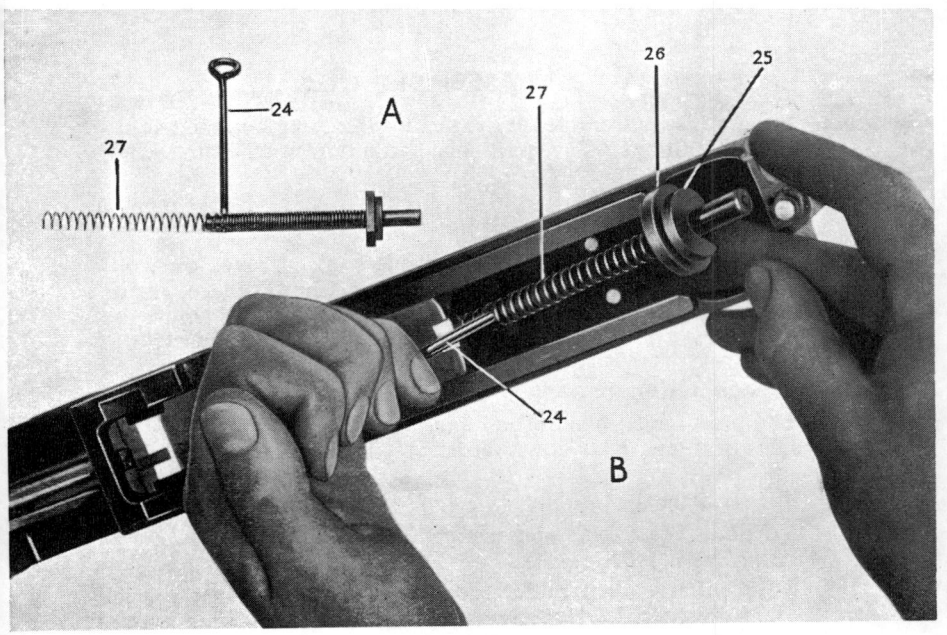

Plate IV

A TOP—*Recoil spring partly compressed on buffer rod, stripping tool inserted (side opposite to flat on buffer collar) ready for reassembling into gun.*

B BOTTOM—*Detail showing method of stripping recoil spring and buffer.*

24	Recoil spring stripping tool.	26	Buffer.
25	Buffer fibre disc.	27	Recoil spring.

TO REASSEMBLE GUN

The gun is reassembled by reversing the forementioned stripping operations. Pay particular attention to the following points :—

H-piece

On one side of the bridge of the H-piece are engraved the word "UP" and an →. When replacing the H-piece in its recessed grooves in the body the word "UP" must be uppermost and the → pointing in direction of the muzzle.

Breech Oiler

The breech oiler should be replaced in the body with the long sides of its felt pads downwards.

Recoil Spring

NOTE.—The recoil spring will be the more easily reassembled by observing the following :—

Hold buffer end of buffer rod against body with flat side of collar downwards. Place recoil spring on buffer rod and compress it a little at a time with each hand. When partly compressed in this manner hold spring in position on rod with left hand and insert stripping tool (handle uppermost) into its hole in buffer rod ; this will retain spring in position. (See Plate IV illustration A.) Replace buffer fibre disc, insert the free end of recoil spring in its hole in the cocking handle. Position rear end of buffer rod in its hole in the rear end of body. (Caution : it will be noted that there is now a gap between the front end of the buffer rod and rear end of bolt, and if the stripping tool were removed, the recoil spring is likely to buckle and fly out through this gap, therefore the following must be observed and carried out.) With the left hand draw back cocking handle until the rear end of bolt just lightly touches the stripping tool. (Now, on withdrawal of the stripping tool, the recoil spring will be guided into its hole in the cocking handle by the buffer rod.) With the left hand withdraw stripping tool.

NOTE.—Recoil spring stripping tool.

Some guns may not be supplied with the stripping tool (24) illustrated in Plates III and IV. It is a simple matter to make one ; the illustrations will serve as a guide for this purpose. Alternatively a nail of appropriate size may be used.

Plate V

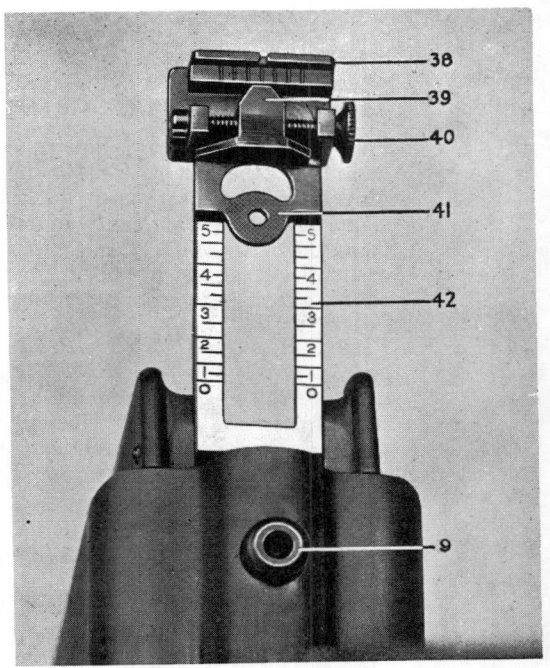

38 Battle (open) sight.

39 Index finger.

40 Aperture sight lateral adjustment screw. (Turned clockwise moves aperture sight to right.)

41 Aperture sight.

42 Leaf engraved for elevation in yards range.

9 Rear end of buffer rod projecting through its seating hole in *body*.

The illustration on this page and the accompanying text describes in detail the sights fitted to the gun. Aiming is done in the same way as with aperture-sighted rifles.

†SIGHTS

Plate V illustrates the backsight with the leaf raised, showing adjustable backsight slide carrying the aperture sight, and the scale graduated on the leaf up to 600 yards. To set sight for elevation, raise the leaf and move the slide to range required. With the leaf lowered, the battle sight (an open backsight) is in position, which is sighted for 50 yards. The aperture sight can be moved laterally by means of a turnscrew; turned in a clockwise direction moves aperture sight to right. When the gun has been zeroed, a line should be placed on index finger to determine the correct adjustment for zero. It is to be noted in this connection that considerable variation in shooting may be obtained, when different firers use the same sight setting; this is due to the different manner in which the gun may be held by them.

NOTE.—For ranges up to 50 yards the leaf sight should be folded down and the slot in the cocking handle used as an open backsight. See text and illustrations on next two pages.

THE USE OF THE COCKING HANDLE IN AIMING

Plate VI opposite shows how to aim using the slot in cocking handle as a backsight. This method of aiming should be employed for ranges up to 50 yards, as a very quick alignment of sights on the target is thereby more quickly accomplished than by using the battle or aperture sight fitted to the gun. Rough alignment of the sights in this manner is sufficiently accurate to enable a man-size target to be hit with certainty up to 50 yards.

PRACTICAL RANGES.—Results to be expected depend a great deal, naturally, on the skill of the gunner, prevailing conditions and type of target. The following may be used as a guide for a shot of average ability. Firing from the shoulder, deliberate aimed single shots up to 100 yds. Aimed single shots, rapid fire up to 50 yds. Quick bursts up to 30 yds. Quick bursts fired from waist up to 20 yds.

AMMUNITION NOTES

.45 Calibre rimless auto-pistol cartridge.

230 grain bullet.   Muzzle velocity approx. 950 f.p.s.

Range	*Height of Bullet above Line of Sight.*
100 yds.	At 50 yds. $4\frac{3}{4}$ inches.
200 yds.	At 50 yds., $16\frac{3}{4}$ inches ; at 100 yds., $22\frac{3}{4}$ inches ; at 150 yds., $18\frac{1}{2}$ inches.

Penetration :   At 25 yds., damp loam soil, approx. 10 inches.

At 25 yds., dry sand, approx. 8 inches.

White Pine, at 50 yds., $5\frac{3}{4}$ inches ; at 100 yds., $5\frac{1}{2}$ inches ; at 200 yds., $4\frac{1}{2}$ inches.

Maximum range at an angle of elevation of 30 degrees is approximately 1,650 yds.

Plate VI. How to Aim using the Cocking Handle
for Quick Alignment of Sights

MAGAZINES

There are two types of magazines as illustrated in Plate VII on the opposite page.

BOX MAGAZINE

This consists of the body, a floor plate and cartridge follower or platform, and spring.

It is important that every care should be taken to prevent the magazine from becoming dented or the lips of the mouth of the magazine from becoming deformed; the latter should be .55 inches apart.

DRUM MAGAZINE

This consists of the main drum casing, the base and cover, a central rotor driven by a mainspring, spiral track cartridge guide and magazine key, as shown in the illustrations.

NOTE.—Great care should be taken to see that magazines are kept clean and dry and their sockets and holes into which the magazine catch engages free from becoming deformed.

FILLING MAGAZINES

TO FILL BOX MAGAZINE

Grasp magazine with left hand, rib toward the body, press cartridges with a downward and backward motion into magazine with right hand. The rounds should be counted as they are loaded into the magazine.

Full capacity, 20 rounds.

NOTE.—Do not fill magazine beyond its capacity of 20 rounds. It is possible to force in an additional round. This should not be done, however, as it needs extra effort on the part of the bolt to feed this round into the breech, with a possibility of a faulty feed or a misfire resulting.

TO EMPTY BOX MAGAZINE

Hold magazine in same manner as for loading. With the base of a cartridge, press out cartridges forward and out of magazine.

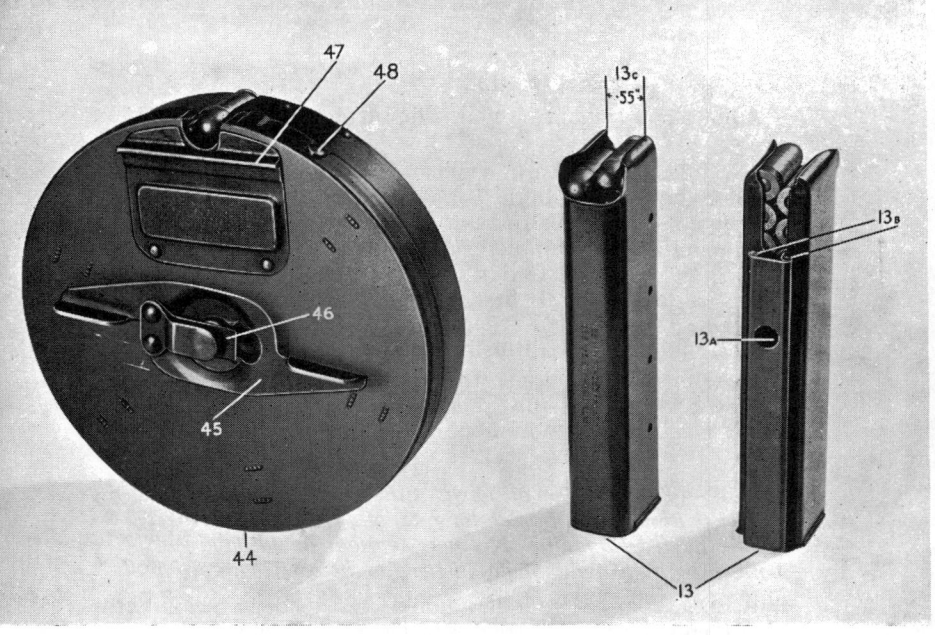

Plate VII MAGAZINES

DRUM MAGAZINE
(Capacity 50 rounds)
44 Drum magazine.
45 Magazine key.
46 Magazine key spring catch.
47 Magazine rib for engage-
    ment with 61, Plate XIV.
48 Cover positioning stud.

BOX MAGAZINE
(Capacity 20 rounds)
13   Box magazine.
13a Hole in magazine rib for
     magazine catch.
13b Magazine ribs for engage-
     ment with 31, Plate XIV.
13c Magazine lips (.55 inch
     apart).

19

TO FILL DRUM MAGAZINE

Lift magazine key spring and slide off magazine key. Remove cover.

Plate VIII clearly shows how the magazine is loaded. The bullets are placed base down in the spiral track. First commence with a complete sector of the rotor at the mouth of the magazine. It will be noted that each sector of the rotor contains five rounds. (See Plate VIII, Illustration C.) The bullets are loaded anti-clockwise, outer spirals first, and should be counted until the magazine is fully loaded—*i.e.*, 50 rounds. Fully loaded magazine is shown in Plate VIII, Illustration D.

Replace magazine cover so that the slot cut in it engages with the cover positioning stud (48, Plate VII). Replace magazine key. Wind magazine key to number of clicks stated on the magazine cover.

NOTE.—The number of clicks stated on magazine gives the correct tension to mainspring for feeding all the 50 cartridges from a fully loaded magazine. No more tension should be given to the mainspring by winding magazine key as the cartridges are fired.

IMPORTANT—No rounds should be loaded beyond the looped end on the spiral track (see shaded portion, Illustration D, Plate VIII), as if this is done when the rotor turns, the round so placed will jam against the looped end of the spiral track.

TO EMPTY DRUM MAGAZINE

With the base of a bullet press out cartridges forward through the mouth of the magazine one by one until magazine is empty.

Alternative Method of emptying Drum Magazine

(See Plate IX.)

Lift magazine key spring and slide off magazine key. Remove cover. All the cartridges in the spiral track, with the exception of those in the first sector of the rotor at the mouth of the magazine, will be found to be loose and can be tipped out by turning the magazine upside down.

Now grasp the rotor with the fingers of the left hand. With the open side of the magazine pointing to the left, hold the rounded side of the magazine against the body and grip the rounded side farthest away from the body with the right hand. Press the side of the magazine against the body and turn the rotor approximately $\frac{1}{4}$ inch anti-clockwise with the left hand, when the

Plate VIII

Drum Magazine Detail

C. Illustration C shows partly loaded drum magazine.

Note.—

    (i) Cartridges placed standing on their bases in body of magazine, bullets up.

    (ii) Loading commenced with a complete section of the rotor loaded 5 rounds, at the mouth of magazine.

    (iii) Loading continued anticlockwise outer spiral, first.

D. Illustration D shows drum magazine fully loaded (to capacity 50 rounds).

45 Magazine key.

49 Magazine cover.

49a Slot in cover for positioning Stud 56a.

50 Magazine cover spiral track guide for bullets.

51 Spiral track guide for cartridges in base of magazine.

52 Cartridge stop.

53 Mouth of magazine.

54 Rotor.

55 Looped end on spiral track (beyond which no cartridge should be placed—*i.e.*, area shown shaded).

56 Base of magazine.

56a Positioning stud for cover slot 49a.

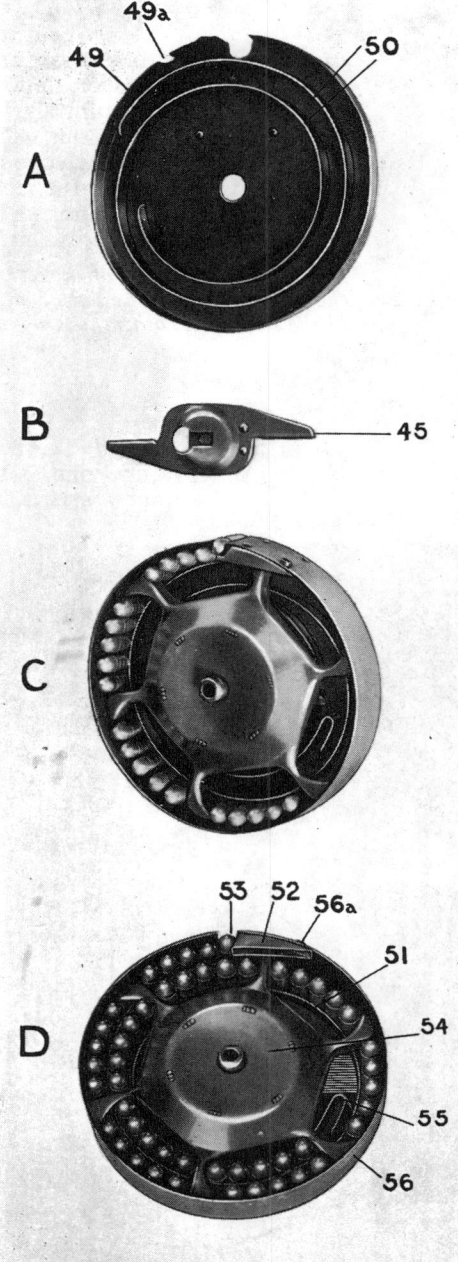

cartridges in the first sector of the magazine can be removed with the right hand or tipped out. The magazine is now empty. (On no account allow the magazine or rotor to spin round uncontrolled, as this may damage or break the magazine mainspring.) Grasp the edge of the magazine again with the right hand; now allow the magazine little by little to unwind round the rotor, controlling its movement by the right hand and pressure against the body, until mainspring is unwound.

NOTE.—The fingers of the left hand should grip the rotor so that the fingers and thumb will be clear of the bullet stop at the mouth of the magazine, otherwise they may be pinched against it.

Plate IX   How to hold Drum Magazine when emptying
by alternative method as described above.

FIRING

BEFORE FIRING

Ensure barrel is clean and dry, and gun is unloaded.
All reciprocating parts must be well oiled.
Make certain that the felt pads of the breech oiler are well oiled.
With change lever set at " Full Auto " and trigger pressed, work cocking handle backwards and forwards several times under control, to ensure that the mechanism is working properly and smoothly, and to distribute the oil evenly.
Magazines must be clean and dry, and correctly loaded.

DURING FIRING

During firing see that the gun is adequately supplied with oil ; all reciprocating surfaces should be oiled frequently and freely so as to ensure smooth and correct functioning of the gun.
If the gun is fired an excessive amount, the residue from the powder gases should be cleaned from the chamber and breech ; this can be done quickly and easily with the breech bristle brush. This ensures smooth and easy extraction of spent cases.

AFTER FIRING

Strip body group (see page 10).
The barrel is cleaned in a similar manner as for a rifle. Use the cleaning rod and barrel cleaning brushes for this purpose. Flannelette patches, approximately 4×4 in., will be found to be correct on the loop end of the cleaning rod. It is important that all surfaces of the gun which are subjected to the effect of gases from the exploded charge should be wiped thoroughly clean and oiled. Make certain the following are thoroughly cleaned :—
The bolt, front end of bolt, extractor, firing pin and spring. (It is important that the latter two parts should be inspected and cleaned after firing, as otherwise they may become impregnated with powder gases and rust unseen in the bolt.)
The inside of the body should be cleaned and also the ejector head. The parts should be thoroughly oiled and the felt pads of the breech oiler saturated with oil when the gun is being reassembled. A good grade of lubricating oil should be used. A drop or two should be given to the trigger mechanism, stop pins, sear and sear lever.
After reassembling the gun a little oil should be given to the rounded front end of the bolt ; the cocking handle should be worked backwards and forwards several times to ensure that the oil is distributed to the reciprocating parts properly.

23

Magazines should be wiped free from powder gases residue with an oiled rag. They should be thoroughly dried when next required for use.

NOTE.—If no cleaning rod available, use pull-through. Flannellette, size 4×8", should be pulled through from breech end until clean, then finally oil with a piece of flannellette, 4×6".

Excessive fouling may be removed by packing out rifle gauze with flannellette so as to fit bore snugly.

TO LOAD GUN

The loading positions are illustrated in the Plates opposite.

(1) Stand facing target.

Grasp pistol grip with right hand, trigger finger outside trigger guard. Butt under right arm, muzzle pointing downwards at 45°. (See Plate X.)

(A) BOX MAGAZINE.

Turn gun to right, grasp magazine in left hand, ribs towards firer, bullets uppermost, insert ribs of magazine in their corresponding vertical grooves in body. (See Plate XI.)

*Push magazine home until magazine catch engages its hole in magazine ribs. Return gun to vertical position. (See Plate XII.) To complete loading cock gun by pulling back with left hand the cocking handle to the rear as far as it will go—*i.e.,* until it clicks twice. (See also Note, page 26.)

(B) DRUM MAGAZINE.

Adopt the same position as shown in Plates X, XI and XII.

Cock gun (the gun must be cocked before a drum magazine can be inserted).

Grasp magazine in left hand, magazine key towards muzzle of gun, mouth of magazine towards barrel. Insert magazine by engaging horizontal ribs of magazine (47, Plate VII) with their corresponding grooves (see 61, Plate XIV) in body.

*Push magazine into position until magazine catch engages its slot in base of magazine.

* NOTE.—No undue pressure or force must be applied when pushing magazine home, as the top of the magazine ribs in the case of the box type, or in the case of the round drum magazine its magazine catch slot, may be damaged. In most cases it will be found best to raise the magazine catch with the thumb of the left hand, while inserting either type of magazine. It is very important to ensure that the magazine catch has properly engaged the magazine catch hole of the box type or slot of the drum magazine.*

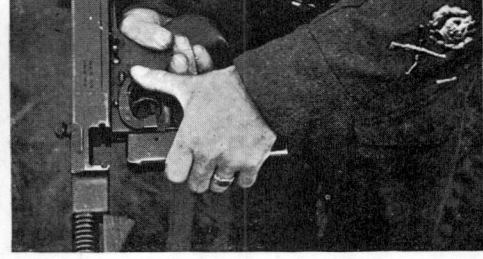

Plate X. Top Left—" Loading Position."

Plate XI. Top Right—Inserting Box Magazine.

Plate XII. Bottom Left—Box Magazine Inserted.

Plate XIII. Use of Thumb of Left Hand to operate Magazine Catch when unloading Gun.

To Unload Gun

Adopt same position as shown in Plate XI. Press magazine catch upwards with the thumb of left hand (see Plate XIII); at the same time withdraw magazine with fingers. Cock gun (if not already cocked), ensure no round is in chamber, hold cocking handle with left hand, press trigger and allow cocking handle to go forward under control. Repeat once.

(7) To Fire Single Shot (Semi-automatic)

Place change lever to " Single." (See 60, Plate XIV opposite. In this illustration fire control lever is shown on " Full Auto.") Place safety catch to " Fire." The gun will now fire shot after shot each time the trigger is pulled, and will stop as soon as the trigger is released or there are no more rounds left in the magazine. Make certain to release the trigger fully between each shot.

(8) To Fire Bursts (Full Automatic)

Place fire control lever to " Full Auto " ; it is shown in this position in Plate XIV.

(9) The gun will now fire shot after shot continuously so long as the trigger is kept pressed .and there are rounds in the magazine. Two to three rounds is the normal number of rounds fired in a burst.

(10) After each burst relay aim and continue firing as desired.

NOTE.—The cocking handle must be in its rearmost position before either the safety catch or the change lever may be moved.

In action the gun is normally carried with a full box-type magazine affixed to the gun, cocking handle forward as shown in Plate XIII, and the safety catch in " Fire " position (it cannot be put at " Safe" with cocking handle forward) and change lever at " Full Auto." As soon as action is imminent it is only necessary to " cock gun " and fire. With a little practice single shots can be quite easily fired with the change lever set at " Full Auto " by a quick tap and release movement on the trigger by the first two joints of the trigger finger. Thus the firer has at his command either single shots or continuous bursts at a movement of the trigger finger without the necessity of moving the change lever.

CAUTION

On no account should the cocking handle be eased forward when a loaded magazine either of the drum or box type is attached to the gun.

The cocking handle must be in its rearmost position before inserting a loaded drum magazine in the gun. A loaded box

26

Plate XIV

4	Ejector.	61	Grooves in body for ribs of drum magazines.
5	*Body.*		
13	Box magazine.	58	Safety catch.
13b	Ribs of box magazine.	31	Grooves in frame for ribs of box magazines.
60	*Change lever.*		
		59	*Magazine catch.*

magazine may, however, be inserted with the breech closed—
i.e., actuator in its foremost position.

(11) When the box magazine is empty the breech remains
open, the cocking handle in its rearmost position ; the empty
magazine may quickly be replaced by either a loaded box
magazine or a loaded drum magazine.

(12) When the drum magazine is empty, and the trigger
is pulled, the bolt will close on an empty chamber, and a
rattling sound will be set up in the drum magazine indicating
it is empty. To reload, pull back cocking handle to its rearmost
position, release magazine catch, remove empty drum magazine
and then either a loaded box magazine or drum magazine may
be inserted.

FIRING POSITIONS

FIRING FROM THE WAIST

In an emergency and when it is required to fire with the utmost rapidity on a surprise target at close quarters, quite effective fire may be delivered by firing the gun from the waist.

Plate XV shows the position firing from the waist. Note particularly :

Left foot advanced with knee bent, weight of body on left leg, right leg braced. The whole attitude one of determination and aggression.

FIRING FROM THE SHOULDER

Plate XVI and the frontispiece show the position firing from the shoulder ; note the right elbow is raised and the right shoulder pushed forward into gun, body and feet same as for firing from the waist. The whole attitude one of determination and aggression.

If the gun is held loosely, shots tend to go high. There is a natural tendency to shoot high with the submachine gun, and when firing bursts the tendency is for the bullets to strike high and to the right. The gunner will soon be able to correct this by practice.

ON THE MOVE

When action is imminent the gun should be carried at the ready, similar to the firing position from the waist. Ordinarily it may be carried at the trail by turning the gun upside down and grasping the pistol grip or slung over the shoulder by means of the sling.

Plate XV   FIRING POSITION FROM THE WAIST

IMMEDIATE ACTION

The gun will stop firing :—

(1) When the box-type magazine is empty, the gun will stop with the cocking handle in its rearmost position. Immediate action to be applied : Change empty magazine for a full one, carry on firing.

(2) When a drum-type magazine is empty, the gun will stop firing with the cocking handle in its foremost position, and a rattling sound will be heard, indicating that the drum magazine is empty.

Immediate action to be applied :—

Cock gun. Replace empty magazine with full magazine ; carry on firing.

(3) Gun still fails to fire :—

Cock gun and give it a sharp flick to the right, when an empty case or cartridge should be thrown out ; if it does not, remove magazine and it will drop out.

NOTE.—It is advisable, if practicable and time permits, to always ensure after carrying out immediate action that the chamber is empty. If care and attention is paid to keeping the gun clean and properly oiled, all magazines in good order, stoppages will seldom occur.

REMINDER NOTES

(1) Do not snap the gun on an empty chamber unnecessarily.

(2) The cocking handle must be in its rearmost position before :

    (A) The safety catch can be put at " Safe."

    (B) The change lever can be moved from " Full Auto " to " Single."

    (C) A drum magazine can be attached to gun.

(3) Never on any account let the cocking handle go forward under control when a loaded magazine is attached to the gun, as a cartridge will be fed into the gun and fired and the cocking handle will fly back and probably hurt the hand or fingers holding the cocking handle.

Plate XVI FIRING POSITION FROM THE SHOULDER

(4) Do not place your finger on the trigger until you have the gun pointing towards the target you wish to hit and are ready to fire.

(5) When gun is not in use, remove magazine, ensure chamber is clear and close bolt. The bolt should not be left cocked as this imposes unnecessary strain on the recoil spring.

(6) Drum magazines should not be left with their springs fully wound, as this imposes unnecessary strain on the springs. They can, if required, be kept fully loaded with cartridges, spring not wound up. It is then a simple matter to wind up each magazine the correct number of clicks before going into action.

PART TWO   **FM 23–40**

BASIC FIELD MANUAL

THOMPSON SUBMACHINE GUN, CALIBER .45, M1928A1

Prepared under direction of the
Chief of Cavalry

UNITED STATES
GOVERNMENT PRINTING OFFICE
WASHINGTON: 1940

WAR DEPARTMENT,
WASHINGTON, *August 19, 1940.*

FM 23–40, Thompson Submachine Gun, Caliber .45, M1928A1, is published for the information and guidance of all concerned.

[A. G. 062.11 (7–2–40).]

BY ORDER OF THE SECRETARY OF WAR:

G. C. MARSHALL,
Chief of Staff.

OFFICIAL:

E. S. ADAMS,
Major General,
The Adjutant General.

TABLE OF CONTENTS

BASIC FIELD MANUAL

THOMPSON SUBMACHINE GUN, CALIBER .45, M1928A1

CHAPTER 1

MECHANICAL TRAINING

SECTION I

DESCRIPTION

■ 1. GENERAL.—The Thompson submachine gun, caliber .45, M1928A1 (fig. 1), is an air cooled, recoil operated, magazine fed weapon weighing about 15¾ pounds with loaded 50-round magazine. The exterior surface of the rear portion of the barrel contains a series of annular flanges which serve to dissipate heat and cool the barrel during firing. The hand of the gunner is protected on the under side of the barrel by a wooden fore grip, and a rear grip is also provided. Sling swivels are attached to the fore grip and stock for attaching the gun sling. The weapon is provided with a "Cutts" recoil compensator to lessen the recoil and the tendency of the muzzle to rise in full automatic fire. By correctly setting the rocker pivot, the weapon may be used for either full automatic or semiautomatic fire. The weapon is fed from a drum type magazine having a capacity of 50 rounds or from a box type magazine having a capacity of 20 rounds.

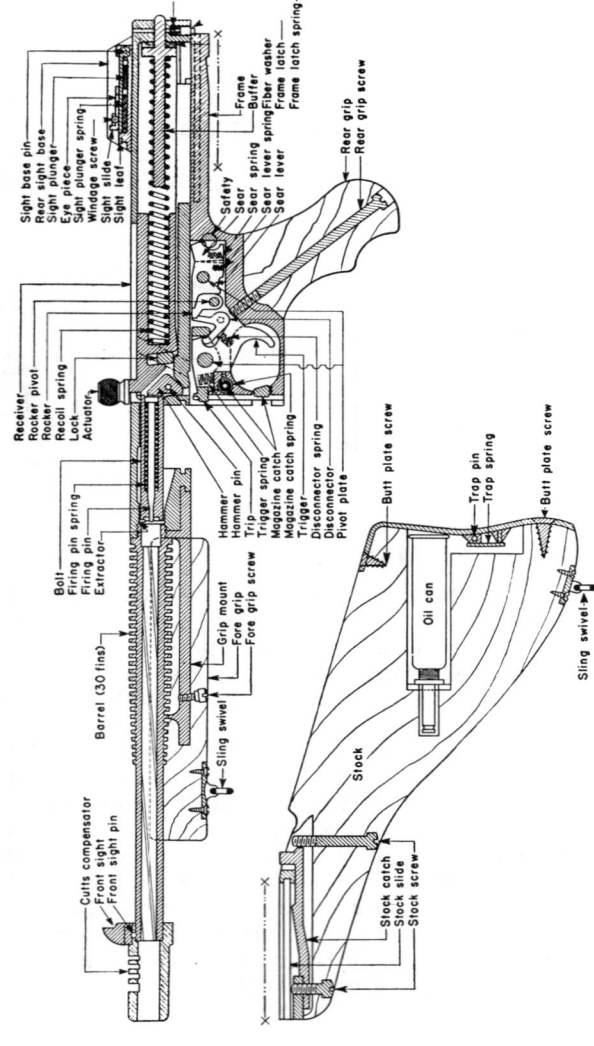

FIGURE 1.—Thompson submachine gun, caliber .45, M1928A1.

■ 2. GENERAL DATA.—*a. Dimension.*

(1) *Barrel.*

Diameter of bore_____inches__ 0.45

Number of grooves_____ 6

Twist in rifling, uniform, right, one turn in_____inches__ 16

Length of barrel_____do____ 10.50

(2) *Gun.*

Over-all length of gun including compensator_____inches__ 33.69

Sight radius_____do____ 22.30

b. Weight.

Gun without magazine_____pounds__ 10.75

Loaded 20-round magazine_____do____ 1.31

Loaded 50-round magazine_____do____ 4.95

Empty 20-round magazine_____do____ .38

Empty 50-round magazine_____do____ 2.63

■ 3. MISCELLANEOUS DATA.

Initial velocity_____feet per second__ 802

Pressure in chamber (approx.) pounds per square inch_____ 12,000–16,000

Weight of ball cartridge (approx.)____grains__ 325

Weight of bullet (approx.)_____do____ 234

Weight of powder charge (approx.)_____do____ 5

Rate of automatic fire (cyclic rate), shots per minute_____ 600–700

■ 4. NOMENCLATURE OF COMPONENT PARTS.

a. Gun (figs. 1 and 2).

1. Actuator, assembly.
2. Barrel.
3. Bolt.
4. Breech oiler (with felt pads).
5. Buffer pilot (with fiber disk and buffer pilot).
6. Disconnector.
7. Disconnector spring.
8. Ejector.
9. Extractor.
10. Firing pin.
11. Firing pin spring.
12. Fore grip.
13. Fore grip screw.
14. Frame.
15. Frame latch.
16. Frame latch spring.
17. Grip mount.
18. Hammer.
19. Hammer pin.
20. Lock.
21. Magazine catch, assembly.

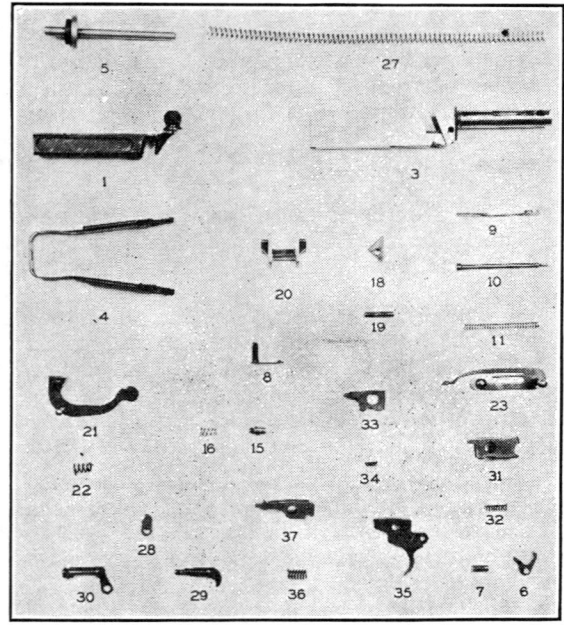

FIGURE 2.—Parts of the gun.

22. Magazine catch spring.
23. Pivot plate, assembly.
24. Rear grip.
25. Rear grip screw.
26. Receiver.
27. Recoil spring.
28. Rocker.
29. Rocker pivot or fire con-
    trol lever.
  b. Box magazine (fig. 3).
    38. Tube, assembly.
    39. Follower.
    40. Spring.
    41. Floor plate.

30. Safety.
31. Sear.
32. Sear spring.
33. Sear lever.
34. Sear lever spring.
35. Trigger.
36. Trigger spring.
37. Trip.

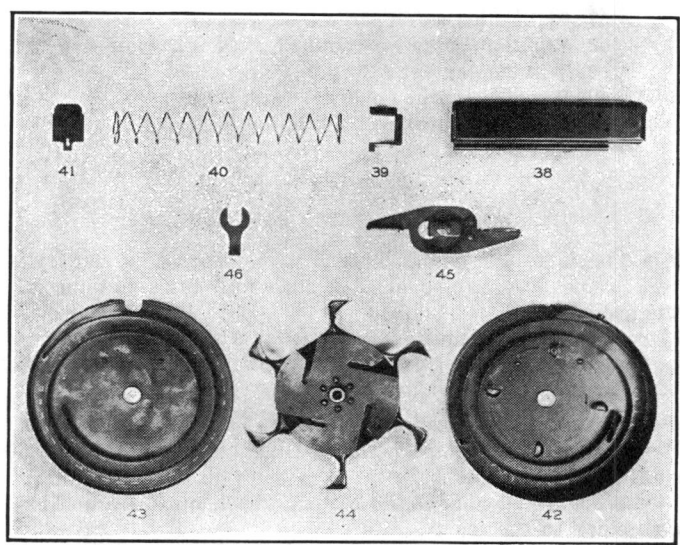

FIGURE 3.—Component parts of the magazines.

c. *Drum magazine* (50 rounds) (fig. 3).
    42. Magazine body, assembly.
    43. Cover, assembly.
    44. Rotor, assembly.
    45. Winding key, assembly.
    46. Rotor retainer.
d. *Sights* (not separately illustrated).
    47. Front sight (assembled to barrel) (fig. 1).
    48. Rear sight (fig. 1).
    49. Eyepiece.
    50. Rear sight base (assembled to receiver).
    51. Sight base pin.
    52. Rear sight leaf plunger (with sight plunger pin).
    53. Sight plunger spring.
    54. Sight leaf (with slide retaining pin).
    55. Sight slide.
    56. Sight slide catch.

57. Sight slide catch screw.
58. Windage screw (assembled), consisting of—
59. Windage screw.
60. Windage screw collar.
61. Windage screw collar pin.

SECTION II

DISASSEMBLING AND ASSEMBLING

■ 5. GENERAL.—*a*. Disassembling is considered under two general headings, removal of groups and disassembling of the groups.

b. A group is a number of components which either function together in the gun or are intimately related to each other and should therefore be considered together.

■ 6. REMOVAL OF GROUPS FROM GUN.—*a*. *Magazine.*—With the bolt in the rearward position, raise magazine catch and slide out magazine.

b. *Stock.*—Press in on butt stock catch button and slide butt stock to the rear.

c. *Frame from receiver.*—Turn safety to "fire" position and rocker pivot to "automatic" or "full auto" position. Pull trigger and allow bolt to go forward gradually by retarding actuator with left hand. Place gun upside down on knee or on table, barrel extending rearward; steady against movement by holding actuator knob. With thumb of left hand depress frame latch and with right hand tap frame, sliding it rearward a short distance. Take gun from table or knee, grasping receiver with left hand; pull trigger and slide frame off to the rear. If the safety is not set at "fire" and the rocker pivot at "automatic" or "full auto," the bolt may be damaged.

d. *Recoil spring.*—Support muzzle of barrel on table or knee with open side of receiver facing operator. Grasp receiver with left hand with thumb in position to engage buffer pilot which projects beyond end of receiver. With thumb of right hand press down on buffer pilot and with thumb of left hand engage flange of buffer. If breech oiler follows, push it back with fingers of right hand. Hold buf-

6

fer and pilot with thumb and forefinger of right hand and withdraw this entire unit from receiver. Care should be taken to obtain a firm hold on spring and buffer pilot to prevent recoil spring, which is compressed, from springing out of operator's hand. When recoil spring at rest measures less than 10½ inches it should be replaced.

e. Bolt, lock, and actuator.—Grasp receiver, bottom up, with left hand. Slide bolt to the rear and lift out. Slide actuator with lock forward and lift out lock. Slide actuator to rear, turn receiver over, and allow actuator to fall out into right hand.

f. Breech oiler.—Press in on the two fingers and lift out. It is not necessary to remove oiler for ordinary cleaning purposes.

■ 7. DISASSEMBLING OF GROUPS.—*a. Magazine, 50 round* (fig. 3).—Remove winding key and lift off cover. Lift up on end of rotor retainer and slide it out of its engagement in the hub. The rotor assembly may then be removed from body. Further disassembly of magazine is not necessary and is prohibited.

b. Magazine, 20-round (fig. 3).—Slide out floor plate. Hold fingers over bottom of magazine tube to keep magazine spring from flying out. Remove follower.

c. Butt stock (fig. 1).—Complete disassembly of butt stock is not necessary for ordinary cleaning. However, when necessary for replacement of broken or worn parts, butt stock slide and butt plate may be removed by removal of screws holding them in place.

d. Frame group (figs. 1 and 2).—(1) Hold frame in left hand, and using back end of actuator (not knob) as a tool in right hand depress short finger of pivot plate and push out rocker pivot with thumb of left hand; lift out rocker and pivot.

(2) Again using actuator but steadying hand with thumb against frame to prevent excessive movement, depress long finger of pivot plate and withdraw safety.

(3) Hold frame upright with grip in right hand. Press simultaneously with both thumbs on sear and trigger pivots. These pivots project sufficiently so that by a quick pressure thereon pivot plate will protrude on the other side far enough

7

to permit withdrawal. While withdrawing pivot plate with left hand, press down on trigger and sear with thumb of right hand to release pressure of springs on pivots. Do not cant pivot plate during withdrawal. The remaining components of the firing mechanism are then free to be removed. Disconnector can be removed from trigger by simply withdrawing it.

(4) To remove magazine catch, rotate it in a counterclockwise direction to its full limit. Removal should be limited to replacement of broken parts. Removal of magazine catch submits magazine catch spring to unnecessary strain and is apt to damage it.

e. Bolt group (figs. 1 and 2).—(1) Drive hammer pin out of bolt from left side; hammer, firing pin, and firing pin spring will then tend to spring out under impulse of firing pin spring. Caution should be exercised to prevent these parts from springing away and becoming lost. Firing pin spring must not be stretched.

(2) Extractor should not be removed for ordinary cleaning or disassembling. To do so submits it to unnecessary strain and is apt to cause it to break or become set.

(3) To remove extractor from bolt insert corner of actuator flange under head of extractor on face of bolt and pull extractor out and up to withdraw it from its groove. When disassembling extractor from or assembling it to bolt, do not lift extractor higher than necessary for lug to clear anchorage hole as otherwise setting or breakage may occur.

f. Receiver group (figs. 1 and 2).—For ordinary training, receiver and parts assembled thereto need not be disassembled.

(1) To remove rear sight leaf, drive out sight base pin and remove leaf. Take care to see that rear sight leaf plunger and spring do not fly out.

(2) Ejector can be removed by lifting leaf sufficiently to disengage detent and unscrewing same from receiver. Do not try to unscrew ejector with bolt assembled and in forward position.

(3) Fore grip can be removed by unscrewing fore grip screw.

(4) Barrel should be removed only for purpose of replacement and then only by authorized ordnance personnel.

■ 8. ASSEMBLING OF GROUPS.—In general, groups and their components are assembled and replaced in the gun in the reverse order of that in which they were removed or disassembled. Certain precautions in assembling (*a* and *b* below) must be observed in order that the parts will function properly after gun is assembled.

a. In assembling trigger mechanism, first see that magazine catch is in place. Assemble springs in their proper recesses. Assemble disconnector to trigger by depressing disconnector spring and sliding disconnector into place.

(1) Place trigger, trip, sear, and sear lever in their respective positions in frame, making sure forward end of sear lever rests on tip of disconnector. To aline these parts, hold frame in left hand and press downward with end of thumb on trigger and base of thumb on sear. Insert pivot plate. To avoid binding, apply gentle pressure with ball of right hand over entire pivot plate.

(2) Insert safety as far as it will go, and using actuator as a tool depress long finger of pivot plate and push safety home. Turn safety to "fire" position.

(3) Place rocker in position in frame with flat side against sear lever. Insert rocker pivot as far as it will go. Using actuator as a tool, depress short finger of pivot plate and push rocker pivot home. Turn rocker pivot to "full auto" position. If rocker is assembled backward, the gun will fire full automatic but not semiautomatic.

b. If extractor has been removed, slide it into place, lifting head only enough to clear stud; avoid excessive pressure. Insert firing pin and spring in bolt being careful to avoid stretching firing pin spring. Place hammer in position with rounded edge upward and push hammer pin into place.

■ 9. REPLACING GROUPS IN GUN.—Before replacing groups in gun, be sure ejector is screwed all the way home and that breech oiler is in place.

a. Place receiver on table or knee, bottom up, and insert actuator in receiver, knob to front. Slide actuator forward and place lock in guideways of receiver, with the word "Up" correctly readable from rear. Slide actuator to rear and place bolt in position.

9

b. Slide bolt forward and start recoil spring, assembled to buffer pilot, into its recess in bolt. Push recoil spring down into bolt until pilot clears end of receiver. Let buffer pilot find its seat in receiver and snap into place.

c. Before fitting frame to receiver, be sure that safety is set at "fire" position and rocker pivot, at "full auto." Slide frame onto receiver and at the same time pull trigger. Frame latch will lock frame in position. Holding trigger depressed, operate actuator back and forth several times to test mechanism.

SECTION III

CARE AND CLEANING

■ 10. GENERAL.—The attention given to a weapon of this type determines largely whether it will shoot accurately and function properly. The bore and chamber must be kept in perfect condition for accurate shooting. Also, it is just as important that the receiver and moving parts be kept clean, lubricated, and in perfect condition for efficient functioning. Care and cleaning will include the magazines, which must be kept free from rust, grit, gum, etc., in order to function properly. When not on the person, the submachine gun should be habitually transported in a suitable boot provided with the necessary brackets for attachment.

■ 11. CLEANING AND LUBRICATING.—*a. General.*—Keep the gun well cleaned and oiled. After each day's firing, clean the bore, chamber, and all parts and surfaces of the receiver, bolt, ejector, and extractor that have been in contact with powder gases. Remove frame from receiver and take bolt out; thoroughly clean front end of bolt and extractor. With the bolt removed, the bolt well, the throat of the receiver, and the ejector head are readily accessible.

b. Bore.—(1) As the barrel of the submachine gun is not removed for cleaning, it must be cleaned from the muzzle, if the submachine gun cleaning rod is used. However, by using the rifle cleaning rod the barrel can be cleaned from the breech. Push rifle cleaning rod through buffer pilot hole in back of receiver and thread a patch through eye of rod.

CAUTION: In cleaning bore, care must be taken not to foul cleaning patch in slots of recoil compensator.

10

(2) Run several wet patches through bore. For this purpose water must be used; warm water is good, and warm soapy water is better. Remove patch, attach cleaning brush, and run brush back and forth through bore several times. Care should be used to insure that brush goes all the way through bore before direction is reversed. Remove brush and run several wet patches through bore. Follow this by dry patches until patches come out clean and dry, then saturate a patch in sperm oil and push it through bore.

NOTE.—Sperm oil (U. S. A. Spec. 2–45A) should be used when available. When not available, motor oil, weight 20, or any light grade machine oil may be used in an emergency.

c. *Chamber.*—The chamber should be cleaned with the chamber cleaning brush at reasonable intervals in extended firing to facilitate extraction of cartridge cases and to prevent pitting and rusting. It is not necessary to disassemble gun for this purpose. The brush is introduced through ejection opening in receiver and should be used vigorously. Upon completion of firing, the brush having been used, the chamber is further cleaned and oiled in the process of cleaning the bore.

d. *Exterior surfaces.*—Wipe off exterior surfaces of gun with a dry cloth to remove dampness, dirt, and perspiration, then wipe off all metal surfaces with an oiled (sperm oil) rag. The stock and grips should be wiped with raw linseed oil.

e. *Magazines.*—It is imperative that magazines be given the best of care and kept in perfect condition. They should be disassembled, wiped clean and dry, and thinly coated with oil. Dirt that gets into them through careless handling during range or other firing must be removed. Care must be exercised in handling magazines to avoid denting or bending, especially the lips of the mouth of the box type magazine.

f. *Lubricating the gun.*—To function efficiently, the gun must be properly lubricated. For this purpose use aircraft machine gun lubricating oil (U. S. A. Spec. 2–27) or sperm oil (U. S. A. Spec. 2–45A).

(1) Having removed frame from receiver, oil should be dropped over pivot points of trigger and trip, sear and sear lever, and disconnector and rocker.

(2) Holding receiver in left hand, open side up, bolt should be slightly drawn back and oil dropped on locking lugs of

lock, on sides of lock, and on all sliding surfaces of bolt and receiver.

(3) *Felt pads in breech oiler should be kept well-saturated with oil.*

(4) After assembling frame to receiver, bolt should be drawn back and a little oil should be dropped on rounded front end of bolt. Actuator knob should be worked back and forth several times to insure penetration of oil to all parts of mechanism.

(5) *All sliding surfaces should be oiled frequently and freely to insure perfect functioning of the gun.*

SECTION IV

FUNCTIONING

■ 12. GENERAL.—The pressure of the gases generated in the barrel by the explosion of the powder in the cartridge is exerted in a forward direction against the bullet, driving it through the bore, and in a rearward direction against the face of the bolt. This force drives the bolt and the actuator together to the rear against the pressure of the recoil spring. During rearward movement, the processes of unlocking, extracting, ejecting empty shells, and compressing the recoil spring are effected; during the forward movement, the processes of feeding, locking, and firing the cartridge are accomplished. To simplify explanation of functioning, the cycle has been divided into three phases as set forth below.

■ 13. BACKWARD MOVEMENT OF RECOILING PARTS (First phase).—The cartridge having been fired, the pressure from the exploding cartridge is transmitted through the forward end of the bolt to the lock and through the lock to the locking surfaces of the receiver. The powder used is fast burning, so that the highest chamber pressure obtained is nearly instantaneous. The lock being made of bronze and the bolt and receiver being made of steel, the high chamber pressure causes the lock to adhere to the locking surface on the receiver, thus locking the bolt in its forward position until this pressure subsides. As soon as the high chamber pressure has subsided, the lock moves upward, clears the locking surfaces in the receiver, and the bolt

12

can move to the rear. The angle of the lock is such that the moment the lock is moved to clear the receiver locking surfaces there is only sufficient powder pressure in the chamber to force the cartridge case and bolt to the rear, eject the empty case, and compress the recoil spring, which thus stores up energy for the forward movement. The empty case is unseated by the chamber pressure as the bolt is unlocked. As soon as the bolt moves back from the abutment on the under side of the receiver, the firing pin spring forces the firing pin to the rear away from the face of the bolt. The empty cartridge is held on the face of the bolt by the extractor. After the bolt has traveled to the rear about 2 inches the ejector, which protrudes in a groove on the left side of the bolt, comes in contact with the base of the empty cartridge and throws it to the right through the ejection opening. The bolt still has about 1¾ inches to go to the rear before the back of the bolt comes in contact with the buffer. The rearward movement of the bolt, carrying the actuator and compressing the recoil spring, expends nearly all the energy imparted by the chamber pressure, so that the bolt does not strike heavily against the buffer. The buffer pad absorbs the remaining shock. On the under side of the bolt there are two sear notches so that, if the bolt strikes the buffer pad, the rear sear notch will pass over the sear and allow the sear to engage in the front notch. If the movement is not strong enough to cause the bolt to strike the buffer pad, the sear will engage in the rear notch. If the bolt moves to the rear far enough to eject the empty cartridge case and to feed the next cartridge from the top of the magazine, the bolt will normally be back far enough to engage the sear with the rear notch.

■ 14. FORWARD MOVEMENT OF RECOILING PARTS (Second phase).—When the trigger is pulled, the bolt moves forward under the action of the recoil spring, carrying the lock and actuator with it. After the bolt moves forward about 1 inch, the forward end of the bolt comes in contact with the back of a cartridge and pushes it forward until the nose of the bullet comes in contact with the bullet ramp in front of the receiver. The lips of the magazine hold the cartridge in a straight line until the cartridge has almost cleared the magazine. The cartridge is guided into the chamber by the bullet ramp and the lips of the magazine. When the cartridge has been seated in the chamber, the extractor snaps around the rim of the cartridge. Just before the bolt reaches its forward position, the lock is cammed down into the locking grooves in the receiver so that the bolt is completely locked as the hammer on the under side of the bolt strikes the receiver. The hammer being of a triangular shape, the lower point strikes the receiver, causing the hammer to pivot around the hammer pin and strike the head of the firing pin with the upper point thereby firing the cartridge. The rectangular surface of the bolt, striking the abutment of the receiver, stops the forward movement.

■ 15. ACTION OF TRIGGER MECHANISM (Third phase).—*a.* (1) *Rocker pivot set at "single."*—When the trigger is pulled, the trigger rotates around the trigger pivot (the forward pin of the pivot plate) and lifts the disconnector up under the sear lever. The sear lever lifts the front end of the sear; this causes the sear to rotate around the sear pivot (the rear pin of the pivot plate), and in so doing depresses the nose of the sear, disengaging it from the sear notch on the under side of the bolt. As the bolt goes forward, the point of the rocker is in the T-groove on the under side of the bolt. When the point of the rocker strikes the rear end of the T-groove, the rocker is forced forward. The rounded part of the rocker comes in contact with the disconnector and forces the disconnector out from under the sear lever. As soon as the disconnector has been disengaged from the sear lever, the sear spring and the sear lever spring force the sear and sear lever up into firing position, so that the sear notch on the bolt will catch on the next rearward movement of the bolt.

(2) *Rocker pivot set at "full auto."*—The rocker pivot is of eccentric design, so that when the rocker pivot is set at "full auto" the rocker is lowered enough to allow the bolt to move forward without striking the point of the rocker. Therefore, the sear remains in its lowered position as long as the trigger is depressed.

b. (1) *Safety set at "fire."*—When the safety is turned toward the front, the flat milled surface is in such a position that the sear is allowed to rotate around the sear pivot.

(2) *Safety set at "safe."*—When the safety is turned toward the rear, the rounded part of the safety engages in a groove in the rear of the sear and locks the sear in its uppermost position. The safety can be turned only when the bolt is to the rear.

c. The magazine catch rotates around its pin and is held down in the engaged position by the magazine catch spring. The stud on the magazine catch is to hold the box type magazine; the drum type is held by the rectangular catch on the left side. The trip functions only when the box type magazine is used. As the magazine empties, a fin on the back of the magazine follower rises up under the trip, causing the trip to rotate around the trigger pin, compressing the disconnector spring, and holding the disconnector forward of the sear lever. Thus the bolt will not go forward on an empty chamber when the box magazine is used.

SECTION V

STOPPAGES AND IMMEDIATE ACTION

■ 16. MISFIRE.—In the event of misfire, retract or cock bolt with a sharp, quick pull on actuator knob. This should insure ejection of misfired cartridge. Inspect chamber to see that it does not contain an unexpended round.

■ 17. OTHER STOPPAGES.—For any other malfunction, retract bolt as above and clear throat and chamber of gun by turning gun over on its side and letting case or cartridge roll out. If necessary, remove magazine and allow cartridge or case to fall out the bottom. While manipulating the gun under these circumstances, always set the gun at "safe."

SECTION VI

SPARE PARTS AND ACCESSORIES

■ 18. SPARE PARTS.—*a.* The parts of any submachine gun will in time become unserviceable through breakage or wear resulting from continuous usage. For this reason spare parts are provided for replacement of the parts most likely to fail, for use in making minor repairs, and in general upkeep of the submachine gun. Sets of spare parts should be maintained as complete as possible at all times and should be kept clean and lightly oiled to prevent rust. Whenever a spare part is used to replace a defective part, the defective part should be repaired or a new one substituted in the spare parts set. Parts that are carried complete should at all times be correctly assembled and ready for immediate insertion in the submachine gun.

b. Twenty- or fifty-round magazines are also issued as spares; the former for use with submachine guns issued for use by cavalry motorcyclists and the latter for submachine guns issued for use by cavalry combat vehicle troops. The quantity of magazines issued per gun is based on the allowance of ammunition authorized. The allowances of spare parts and of magazines are prescribed in SNL A–32.

■ 19. ACCESSORIES.—a. *General.*—Accessories include the tools required for disassembling and assembling and for the cleaning and preservation of the gun. They must not be used for any purpose other than as prescribed. There are a number of accessories, the names or general characteristics of which indicate their uses or applications. Therefore, detailed description or method of use of such items is not outlined herein. However, accessories embodying special features or having special uses are described in b below.

b. (1) *Brush, chamber cleaning, M6.*—The brush consists of a steel wire core with bristles, the core being twisted in a spiral to hold the bristles in place. It is used to clean the chamber of the submachine guns.

(2) *Brush, cleaning, caliber .45, M5.*—The brush consists of a brass wire core with bristles and tip. The core is twisted in a spiral and holds the bronze bristles in place. The brass tip which is threaded for attaching the brush to the cleaning rod is soldered to the end of the core.

(3) *Case, accessory and spare parts, M1918.*—This is a leather box-shaped case, approximately 2¼ inches wide, 3½ inches high, and 5½ inches long. It is used to carry spare parts and a number of the smaller accessories.

(4) *Rod, cleaning, submachine gun.*—This consists of a long steel rod having a circular loop at one end and a clean-

ing patch slot at the other end. Permanently affixed to the cleaning patch end of the rod is a head having a threaded hole to receive the cleaning brush, caliber .45, M5.

(5) *Sling, gun, M1923 (webbing).*—The gun sling is fastened to the swivels provided on the gun. It consists of a long and short strap, either of which may be lengthened or shortened as desired to suit the particular soldier using it.

(6) *Thong.*—The thong consists of a tip with cleaning patch slot and a weight tied to the ends of a 30-inch length of cord. It is used in cleaning the bore of the submachine gun.

SECTION VII

INDIVIDUAL SAFETY PRECAUTIONS

■ 20. GENERAL RULES.—*a.* Before firing—

(1) Test trigger mechanism at "safe" and "single."

(2) See that bore is clear and clean.

(3) Work bolt back and forth rapidly several times to see that it is clean, well-oiled, and works freely.

(4) Examine magazines and eliminate faulty ones.

(5) See that each magazine is free from dirt and that it is properly loaded.

b. For range practice, insert loaded magazine only on order of the officer in charge of firing. Do not attach loaded magazine until ready to fire.

c. Carry gun with bolt retracted and safety on "safe" until ordered to attach magazine, and keep safety on "safe" until gun is raised to fire.

d. Keep trigger finger outside trigger guard until gun is raised to fire.

e. From time magazine is attached until gun is cleared and clearance checked, keep gun pointed toward target, whether firing dismounted or from a vehicle.

f. For semiautomatic fire, make certain that rocker pivot is set on "single" and the safety on "safe" before attaching magazine.

g. For full automatic fire, make certain that rocker pivot is set on "full auto" and the safety on "safe" before attaching magazine.

17

h. For vehicular firing at moving ground targets, the gunner must keep the safety on "safe" at all times when his vehicle is moving.

i. Habitually set the safety at "safe" while changing magazines and during lulls in firing.

j. To clear gun, first remove the magazine.

k. At CEASE FIRING and upon halting at the finish of a vehicular run, set the safety at "safe," remove magazine, and see that no cartridge remains in the chamber before turning away from the firing point.

l. After gun has been cleared and checked for clearance, close the bolt on an empty chamber.

SECTION VIII

AMMUNITION

■ 21. GENERAL.—The information in this section pertaining to the several types of cartridges authorized for use in the Thompson submachine gun, caliber .45, M1928A1, includes a description of the cartridges, means of identification, care, use, and ballistic data.

■ 22. CLASSIFICATION.—Based upon use, the principal classifications of ammunition for this rifle are—

> Ball, for use against personnel and light matériel targets.
>
> Tracer, for observation of fire and incendiary purposes.
>
> Dummy, for training (cartridges are inert).

■ 23. LOT NUMBER.—When ammunition is manufactured an ammunition lot number, which becomes an essential part of the marking, is assigned in accordance with specifications. This lot number is marked on all packing containers and the identification card inclosed in each packing box. It is required for all purposes of record, including grading and use, reports on condition, functioning, and accidents in which the ammunition might be involved.   Since it is impracticable to mark the ammunition lot number on each individual cartridge, every effort should be made to maintain the ammunition lot number with the cartridges once the cartridges are removed from their original packing.   Cartridges which have

18

been removed from the original packing, and on which the ammunition lot number has been lost, are placed in grade 3. It is therefore necessary when cartridges are removed from original packings that they be so marked that the ammunition lot number is preserved.

■ 24 GRADE.—Current grades of existing lots of small arms ammunition are established by the Chief of Ordnance and are published in Ordnance Field Service Bulletin 3–5. No lot other than that of current grade appropriate for the weapon will be fired. *Grade 3 ammunition is unserviceable and will not be fired.*

■ 25. IDENTIFICATION.—*a. Markings.*—The contents of original boxes are readily identified by the markings on the box. Similar markings on the carton label identify the contents of each carton.

b. Color bands.—Color bands painted on the sides and ends of the packing boxes further identify the various types of ammunition. The following color bands for cartridges are used:

      Ball_____ Red.
      Tracer _____ Green on yellow.
      Dummy _____ Green.

c. Types and models.—(1) The cartridges authorized for use in this weapon are designated—

      Ball, caliber .45, M1911.
      Tracer, caliber .45, M1.
      Dummy, caliber .45, M1921.

(2) When removed from their original packing containers, the cartridges may be identified except as to ammunition lot number and grade by physical characteristics described below:

(a) *Ball.*—The bullet of the ball cartridge has a gilding metal jacket.

(b) *Tracer.*—The bullet of the tracer cartridge has a gilding metal jacket which is painted red for a distance of approximately $\frac{3}{16}$ inch from the tip.

(c) *Dummy.*—The dummy cartridge is identified by its cartridge case which is tinned and has a ⅛-inch hole in the body. The bullet is the same as that used in the ball cartridge.

19

■ 26. CARE, HANDLING, AND PRESERVATION.—*a.* Small arms ammunition is not dangerous to handle. Care, however, must be exercised to keep the boxes from becoming broken or damaged. All broken boxes must be immediately repaired and all original markings transferred to the new parts of the box. The metal liner should be air tested and sealed if equipment for this work is available.

b. Ammunition boxes should not be opened until the ammunition is required for use. Ammunition removed from the airtight container, particularly in damp climates, is apt to corrode thereby causing the ammunition to become unserviceable.

c. The ammunition should be protected from mud, sand, dirt, and water. If it gets wet or dirty wipe it off at once. Light corrosion, if it forms on cartridges, should be wiped off. However, cartridges should not be polished to make them look better or brighter.

d. No caliber .45 ammunition will be fired until it has been identified by ammunition lot number and grade.

e. Do not allow the ammunition to be exposed to the direct rays of the sun for any length of time. This is liable seriously to affect its firing qualities.

■ 27. STORAGE.—Whenever practicable, small arms ammunition should be stored under cover. Should it be necessary to leave small arms ammunition in the open, it should be raised on dunnage at least 6 inches from the ground and the pile covered with a double thickness of paulin. Suitable trenches should be dug to prevent water from flowing under the pile.

■ 28. BALLISTIC DATA.—Approximate maximum ranges are as follows:

Yards

Cartridge, ball, caliber .45, M1911_____ 1, 600
Cartridge, tracer, caliber .45, M1 (estimate only)_ 1, 600

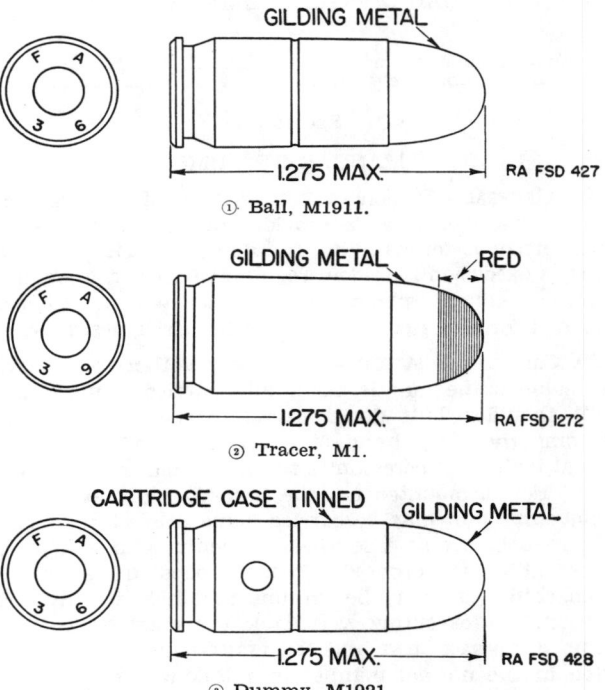

① Ball, M1911.

② Tracer, M1.

③ Dummy, M1921.

FIGURE 4.—Cartridges, caliber .45.

CHAPTER 2

MANUAL OF THE SUBMACHINE GUN

SECTION I

MANUAL OF ARMS

■ 29. GENERAL.—In general, the manual of the submachine gun is prescribed so as to provide uniform, simple, safe, and quick methods for handling the gun. Precision and simultaneous execution is seldom required; however, a simple and effective manual which can be executed in cadence is included for those occasions when its use is desirable.

■ 30. CARRYING POSITION.—Except when otherwise prescribed, the submachine gun is habitually carried slung over the right shoulder, butt down, barrel to the rear, right hand grasping the sling, hand in front of armpit (fig. 5). For formal drills and ceremonies the box magazine is habitually used. For dismounted marches and for field exercises, the submachine gun may be carried slung over either shoulder. When troops are at ease, the submachine gun is kept slung unless otherwise ordered. When troops are at rest, the submachine gun may be unslung and held in any desired position. In executing ATTENTION the carrying position is assumed. PARADE REST and the RIGHT-HAND SALUTE are executed in the normal manner by releasing the grasp of the right hand from the sling.

FIGURE 5.—Carrying position.

■ 31. POSITION OF RAISE ARMS (fig. 6).—Executed at the command: 1. RAISE, 2. ARMS, or when other members of the command are executing the manual of the pistol at the command RAISE PISTOL. This position is assumed by the gunner from the position of attention by grasping the fore grip with the left hand, withdrawing the right arm from between the gun and the sling, then assuming the position as shown in figure 6. The gunner standing at attention grasps the rear grip with his right hand, forefinger extended along the outside of the trigger guard; holds the gun with the butt plate resting on his belt over his right hip, barrel extending upward to the front in a vertical plane and at an angle of 45°. He steadies the gun by pressing the stock against his right side with his right forearm. Left arm and hand are at the left side in a natural unconstrained position.

■ 32. POSITION OF PORT ARMS (fig. 7).—Executed at the command: 1. PORT, 2. ARMS. The gunner standing at attention holds the piece in a vertical plane parallel to and about 4 inches in front of his body, barrel extending upward to the left at an angle of 45°. The right hand grasps the small of the stock. The left hand holds the fore grip and is opposite to and at the same level as the point of the left shoulder.

23

FIGURE 6.—Position of RAISE ARMS.

FIGURE 7.—Position of PORT ARMS.

■ 33. POSITION OF INSPECTION ARMS (fig. 8).—Executed at the command: 1. INSPECTION, 2. ARMS, or when other members of the command are executing the manual of the pistol, at the command INSPECTION PISTOL. The position is the same as PORT except that the actuator has been pulled to the rear, opening the belt, and the safety has been set at safe.

FIGURE 8.—Position of INSPECTION ARMS.

■ 34. POSITION OF PRESENT ARMS (fig. 9).—Executed at the command: 1. PRESENT, 2. ARMS. The gunner, standing at attention and grasping the small of the stock lightly with the right hand and the fore grip with the left hand, holds the piece 4 inches in front of the center of his body in such

FIGURE 9.—Position of PRESENT ARMS.

manner that the barrel is vertical and to the rear with the muzzle up. The gunner's right arm should be straight without constraint.

■ 35. EXECUTION OF THE MANUAL.—In describing the execution of the following movements, it is assumed that neither the box nor the drum magazine has been attached to the gun. If desired, the manual can be executed while either magazine is attached, with minor and obvious modifications.

a. RAISE ARMS to PORT ARMS (in two counts).—(1) Carry the gun diagonally across the front of the body so that the barrel and the muzzle are up; at the same time grasp the fore grip smartly with the left hand.

(2) Move the right hand from the rear grip to the small of the stock.

b. PORT ARMS to RAISE ARMS (in three counts).—(1) Remove the right hand from the small of the stock and grasp the rear grip, with the forefinger extending along the outside of the trigger guard.

(2) With the left hand still grasping the fore grip, carry the piece to the position of RAISE.

(3) Drop the left hand smartly to the side.

c. RAISE ARMS to INSPECTION ARMS (in three counts).—(1) Raise the left hand and pull the actuator to the rear, cocking the piece. Set the safety at safe using the thumb of the right hand.

(2) Execute the first movement of RAISE ARMS to PORT ARMS.

(3) Execute the second movement of RAISE ARMS to PORT ARMS.

d. INSPECTION ARMS to RAISE ARMS (in four counts).—(1) Execute the first movement of PORT ARMS to RAISE ARMS.

(2) Execute the second movement of PORT ARMS to RAISE ARMS.

(3) With the right thumb move the safety to the fire position, and with the right forefinger pull back on the trigger; the left hand moves to the actuator, grasps it, and allows the bolt to move forward without shock.

(4) Execute the third movement of PORT ARMS to RAISE ARMS.

e. RAISE ARMS to PRESENT ARMS (in three counts).—(1) Execute the first movement of RAISE ARMS to PORT ARMS.

26

(2) Execute the second movement of RAISE ARMS to PORT ARMS.

(3) Carry the piece smartly to the position of PRESENT ARMS.

f. PRESENT ARMS to RAISE ARMS (in four counts).—(1) Carry the piece smartly to the position of PORT ARMS.

(2) Execute the first movement of PORT ARMS to RAISE ARMS.

(3) Execute the second movement of PORT ARMS to RAISE ARMS.

(4) Execute the third movement of PORT ARMS to RAISE ARMS.

SECTION II

LOADING AND FIRING

■ 36. LOADING MAGAZINES.—*a. Box magazines.*—(1) The normal capacity of a box magazine is 20 cartridges. The cartridges feed into the magazine easily. If for any reason excessive force is required to feed the cartridges out of the magazine, the energy of the bolt is taxed to such an extent that a misfire may result. The forward edge of the magazine is rounded to prevent loading cartridges backward.

(2) The lips of the mouth of the box magazine should be a distance of .55 inch apart. If by accident the magazine mouth should become deformed, the lips should be carefully bent back to this dimension.

b. Drum magazines.—To load a drum magazine, remove the winding key by lifting the flat spring thereon and sliding the key off. The cover can then be removed. Place the cartridges, bullet up, in the spiral track of the body, beginning with a full section at the mouth. The simplest method to begin loading is to fill one outer section and then rotate the rotor until this section reaches the mouth. Thereafter, continue to fill successive sections until the end of the spiral track has been reached. Fill each section completely; do not skip any section and do not fill beyond the end of the spiral track. After the magazine is properly filled, replace the cover and key, and wind to the number of clicks indicated on the magazine name plate. Drum magazines, when wound to the number of clicks indicated on the case, should not be rewound after shots have been fired, as the resultant strong spring tension interferes with the surety of action of the

27

gun, as well as incurring the possibility of breaking the main spring of the magazine.

■ 37. Loading the Weapon.—At the command INSERT LOADED MAGAZINE, the bolt is pulled to the rear position, the safety is turned to safe, and the magazine of the type desired placed in the grooves of the gun. The box type magazine is pushed upward into the groove at the end of the trigger guard until it clicks into place (fig. 10). The 50-round drum magazine is inserted from left to right in the horizontal grooves of the piece (fig. 11). The magazine should always be pushed well

FIGURE 10.—Inserting box magazine.

up or in so that the magazine catch can snap into position and hold the magazine securely. In order to fire, the safety is turned to "fire." When the trigger is pulled, the forward movement of the bolt feeds a round of ammunition from the magazine into the chamber and fires the gun.

■ 38. Firing.—a. In single shot (semiautomatic) fire, release the trigger quickly after each shot to re-engage the sear, as each pull of the trigger releases but a single shot when the fire control is set for "single."

b. In full automatic fire, release the trigger after each burst of two or three shots and realine sights on the target before firing again. Short bursts can be more accurately placed than long bursts.

c. In firing from the box (20-round) magazine, the bolt is automatically held in the open position when the magazine has been emptied. To close the bolt on an empty chamber, remove magazine and let the bolt go forward slowly, retarding the actuator with the left hand. Do not snap the bolt on an empty chamber; ease it gradually into the forward position.

d. In firing the drum (50-round) magazine, the bolt automatically closes on the empty chamber when the magazine has been emptied.

FIGURE 11.—Inserting drum magazine.

e. This gun fires on the forward stroke of the bolt. Unless the magazine is removed before the bolt is released by pulling the trigger, the gun will continue to load itself as long as there is ammunition in the magazine.

f. Correct firing positions for the submachine gun differ from those prescribed for the rifle in that, with the submachine gun, the gunner's right shoulder is shoved strongly forward against the butt of the weapon, and his body is so placed that he faces more directly toward his target than does the rifleman. It is particularly important for quick and accurate shooting with the submachine gun to assume the correct position whenever full automatic fire is used.

CHAPTER 3

MARKSMANSHIP, KNOWN DISTANCE TARGETS

SECTION I

PREPARATORY TRAINING

■ 39. PURPOSE.—The purpose of preparatory training in submachine gun marksmanship is to teach the gunner the essentials of good shooting and to develop fixed and correct shooting habits before he undertakes range practice.

■ 40. FUNDAMENTALS.—To become a good submachine gunner a soldier must be thoroughly trained in the following essentials of good shooting:

Correct sighting and aiming.

Correct range estimation.

Correct positions.

■ 41. PHASES OF TRAINING.—*a.* Marksmanship training is divided into the following phases:

(1) Nomenclature, functioning, and care of the weapon.

(2) Preparatory marksmanship training.

(3) Range practice.

b. The submachine gunner will be made proficient in mechanical training before he receives instruction in marksmanship training.

c. A thorough course in preparatory training precedes any range practice. This preparatory training is given to all soldiers expected to fire the submachine gun during range practice, including those previously qualified.

d. All units armed with the submachine gun will include instruction in mechanical training and fundamental elements of submachine gun marksmanship—sighting, aiming, range estimation, and positions—for all recruits. Instruction commences with the initial instruction of the recruit, in conjunction with other arms and weapons, and continues throughout the period of recruit training.

■ 42. METHODS OF TRAINING.—To insure proper instruction, well-qualified officers and noncommissioned officers are placed in charge of instruction. Instruction must be given by subject in the proper sequence so that the student will employ what he has been taught when he moves to the next step or phase of training. Brief talks followed by demonstrations by the instructors precede any work by the students. No student is progressed from one stage of training to the next stage unless he is qualified. Constant supervision by noncommissioned and commissioned officers, with individual instruction where and when necessary, is essential. The coach and pupil system of instruction is used whenever a man is in the firing position. Instruction and exercises are rotated in order to keep up interest and enthusiasm in the class.

■ 43. EQUIPMENT.—*a.* The equipment required for preparatory marksmanship training is simple and readily improvised from materials at hand. The following list includes the necessary equipment:

 Submachine gun.
 Sighting disk.
 Submachine gun rest.
 Material for blackening sights.
 Pencil.
 Paper.
 Targets E, F, and M.
 Cleaning and preserving materials.

b. The equipment required for range practice includes, in addition to the necessary items listed above, the materials necessary to construct and maintain target ranges (see sec.

IV, ch. 3 and ch. 5), targets, ranges, weapons, vehicles, and range safety equipment.

■ **44. USE OF SIGHTS.**—*a. General.*—(1) Sights are set by raising the leaf and sliding the slide to the range desired. Lateral correction is obtained by turning the small thumb screw.

(2) The submachine gun is sighted for a "fine sight," that is, with the top of the front sight just showing in the bottom of the peep or open sight. The personal element of holding the submachine gun affects the accuracy of fire, especially when on automatic. Consequently there may be considerable variation in hits when the same sight setting is used by different persons. Holding the gun loosely tends toward shooting high.

b. Windage and drift.—One point of windage will change the point of strike 1 foot at 100 yards. At ranges of 300 yards or more, lateral correction is also made for drift. The drift table showing drift to right from line of bore with 230-grain bullet is as follows:

Range in yards__	50	100	150	200	250	300	350	400	450	500
Drift in feet_____	0	0.10	0.25	0.50	0.85	1.40	2.30	3.25	4.50	5.90

NOTE.—Drift at ranges of less than 300 yards is less than 1 foot and therefore negligible.

c. Range.—(1) The weapon is primarily intended for firing at short ranges where quick shooting is required. Therefore the battle sight is normally used. For firing with the battle sight, the sight leaf is laid down and the open sight used. No windage or drift correction is necessary.

(2) The trajectory of the submachine gun bullet is less flat than that of a rifle bullet. It is important that the gunner be taught to allow for this characteristic in aiming and to understand the effect of the trajectory on his line of sight. The following table gives the height of the trajectory at points along the trajectory when the sight is set at various ranges:

HEIGHT OF TRAJECTORY ABOVE LINE OF SIGHT FOR STANDARD AMMUNITION

[Height of trajectory at points indicated]

Range (yards)	50		100		150		200		250	
	Feet	Inches	Feet	Inches	Feet	Inches	Feet	Inches	Feet	Inches
100	0	4¹³⁄₁₆		0						
200	1	4¹³⁄₁₆	1	10¹³⁄₁₆	1	6⅝		0		
300	2	6⅝	4	3	5	0	4	7³⁄₁₆	2	1⅜
400	3	11⅜	7	0	9	2⅜	10	1⁷⁄₁₆	9	9⅝
500	5	7	10	2⅜	14	0	16	6	17	8⅜

Range (yards)	300		350		400		450		500	
	Feet	Inches	Feet	Inches	Feet	Inches	Feet	Inches	Feet	Inches
100										
200										
300		0								
400	8	2⅜	4	10¹³⁄₁₆		0				
500	17	7³⁄₁₆	15	10⁵⁄₁₆	12	6	7	1¹³⁄₁₆		0

■ **45. SIGHTING AND AIMING.—*a. General.*—**In all preparatory exercises involving sighting and aiming and in all range firing, both sights of the rifle are blackened. The blackening is done by holding each sight for a few seconds in the point of a small flame which is of such a nature that a uniform coating of lamp black will be deposited on the metal. Materials most commonly used for this purpose are carbide lamp, kerosene lamp, candles, shoe paste, and stove polish.

b. Exercises.—(1) *No. 1.*—(*a*) *Purposes.*—To show the correct alining of sights.

(*b*) *Method.*—The instructor places the submachine gun in the submachine gun rest and alines the rest with a blank sheet of paper. The instructor then alines the sighting disk with the sights of the submachine gun by directions to the marker who controls the disk. When the disk is correctly alined the instructor commands: HOLD. The coach moves away from the submachine gun and directs the pupil to look through the sights in order to observe the correct aim.

(2) *No. 2.*—(*a*) *Purpose.*—To show the importance of uniform and correct aiming.

(*b*) *Method.*—The submachine gun with the sights blackened is placed in the rest and pointed at a sheet of paper mounted on a box about 50 feet away. The pupil takes position behind the weapon and looks through the sights without touching the weapon or the rest. The pupil directs the marker with the small disk to move the disk until the sights are correctly alined with the bottom of the disk and then commands: MARK. The marker without moving the disk makes a small dot on the paper through a hole in the center of the disk. The marker then moves the disk to change the alinement. The same process is repeated until three dots are made. These dots outline the shot group. The instructor should discuss with the student the size and shape of the shot group, pointing out the errors.

(3) *No. 3.*—(*a*) *Purpose.*—To demonstrate how changes in sight setting move the shot group.

(*b*) *Method.*—The same procedure is used as explained in the second exercise to obtain the initial shot group. When this has been accomplished, the instructor directs the student to change the sights in range by a certain number of yards and to make a second shot group. Changing the sights is accomplished carefully to avoid moving the submachine gun in the rest. After the sight change has been made, a second shot group is made by the student. The instructor then directs that the sight be changed on the windage scale by a certain number of points. After this change has been accomplished, a third shot group is made. On completion of the exercise the instructor and pupil discuss the size, shape, and relative positions of each shot group and review the principles of sight changes.

■ 46. RANGE ESTIMATION.—*a. General.*—(1) The submachine gunner must be well trained in hasty range estimation and its application to markmanship. Because the weapon is normally employed quickly and at short ranges, the following methods of range estimation are used:

Estimation by eye.

Observation of fire.

(2) The usual method of range estimation is by eye. The submachine gunner is taught to estimate accurately and fix permanently in his mind two distances, 50 yards and 100 yards. Targets at other ranges are estimated in comparison with these units of measure.

(3) When the effect of a shot or bursts of shots can be seen by the gunner, he corrects the range setting applied by estimation in order to increase the accuracy of his fire.

b. Exercises.—The following exercises can be used as guides in instructing the submachine gunner in range estimation. Ranges used are short and at no time greater than 500 yards. The exercises are especially suitable for class exercises.

(1) *No. 1.*—(*a*) *Purpose.*—To familiarize the gunner with the units of measure, 50 yards and 100 yards.

(*b*) *Method.*—The units of measure, 50 yards and 100 yards, are staked out on the ground up to 500 yards. The gunner is required to become familiar with the appearance of the unit of measure from the prone, kneeling, and standing positions on the ground and from a moving vehicle in his normal riding position.

(2) *No. 2.*—(*a*) *Purpose.*—To give practice in range estimation.

(*b*) *Method.*—From a suitable point, ranges are previously measured to normal targets within 500 yards. The gunner is required to estimate the ranges to the various objects as they are pointed out by the instructor and record his estimation on a sheet of paper. At least one-half of the estimates are made from the kneeling and sitting positions. Thirty seconds is allowed for each estimate. When all the ranges have been estimated, the paper is checked by the instructor and the true ranges given to the student.

■ **47. POSITIONS.**—*a. General.*—Assuming a correct position in firing, the submachine gun has direct bearing on the effectiveness of fire. The positions from which the submachine gun may be fired are standing, kneeling, prone, and from the hip while marching. The latter is relatively ineffective and should rarely be used. Each gunner is given sufficient practice to enable him to assume all positions rapidly and efficiently.

FIGURE 12.—Standing position.

  b. *Standing* (fig. 12).—The standing position is the position normally used. The left foot is well advanced with the body leaning slightly forward, about two-thirds of its weight supported by the left foot. The right foot is firmly planted on the ground in the approximate walking position. The left elbow is under the gun and the right elbow raised to shoulder height. The cheek is placed firmly against the stock. The trunk of the body is twisted so as to shove the right shoulder forward strongly against the butt of the piece in the direction of fire. The right shoulder must be tense and pushed into the gun by turning the trunk at the hips and humping the shoulder. The recoil is very slight for any one shot, but the rapid accumulation of the successive slight recoils in full automatic fire tends to push the gunner's shoulder backward, if he is not well braced for the thrust, and the muzzle of his gun will move upward and to the right, thus affecting the accuracy of his fire. If the proper position is taken, the gun will not climb and need not be held down with the left hand.
  c. *Kneeling* (fig. 13).—The kneeling position affords a steadier aim than does the standing position and is useful

FIGURE 13.—Kneeling position.

whenever the gunner can crouch behind a rock, log, or other protection. In the kneeling position, the left toe points at the target, the lower leg being practically vertical. The right knee is on the ground, pointing about 45° to the right of the target. In general, this position is similar to the kneeling position prescribed for a rifleman, except that the submachine gunner is facing more directly toward his target so as to force his right shoulder forward against the butt of his gun. The gunner supports most of the weight of the gun with his left hand; it is not necessary to grasp the piece tightly with the left hand. The gunner's right elbow is raised to the height of his shoulder as in the standing position.

d. *Prone* (fig. 14).—The prone position is the steadiest one; it should be used whenever time and terrain permit, particularly for firing at ranges over 100 yards. This position is similar to the prone position prescribed for a rifleman, except that the submachine gunner is facing more directly toward his target so as to force his right shoulder forward against the butt of his gun.

FIGURE 14.—Prone position.

■ 48. MARKSMANSHIP EXERCISES.—a. First.—(1) *Purpose.*—
This exercise is designed as a preliminary step in teaching the
gunner to pick up his target by any slight movement in his
general field of vision, and to move the gun rapidly from a
standing position to the target with sights properly alined.
The gunner simulates firing a shot or two at each of several
targets which are exposed in rapid succession. In combat he
is usually required to fire on several point targets in rapid
succession; he must aim and fire quickly and accurately.

(2) *Method.*—Place the gunner in the standing position at
A (fig. 15), gun cocked, unloaded, and set for semiautomatic
fire; targets (sec. IV) are partially concealed and all turned
with edges toward the gunner. The gunner is instructed to
look generally over his field of fire by shifting his eyes slightly,

without focusing his eyes on any particular object, until he sees one of the targets move. The target is exposed on the order of an instructor who gives the gunner no advance indication of which target is to be used. The gunner brings his line of aim on this target at the proper height and simulates firing one shot. Without moving the gun from his shoulder, the gunner looks generally over his field of fire until he sees another target move and then simulates firing a shot at that target. Targets initially remain exposed about 5 seconds; this is later reduced to 3 seconds. In a short time, a gunner will be able to pick up the movement of a target and simulate fire on it with facility. When this has been accomplished, the gunner moves forward about 10 yards, sets the gun for full automatic fire, and simulates firing short bursts of 2 to 4 shots on each target of groups C and D. These targets initially remain exposed for 10 to 15 seconds; this later is reduced to 5 seconds.

   b. Second.—After the exercise described in *a* above has been mastered, the gunner goes through the dismounted practice course as prescribed in section II, using no ammunition. This is repeated until he can go through the entire course with facility in the allotted time.

   c. Third.—After the gunner has become proficient in the two exercises described above, he is then practiced in the vehicular courses (figs. 19 and 20) without ammunition. This exercise may be postponed until after he has actually fired the dismounted practice course. In any event, he repeatedly practices the vehicular courses before being permitted to fire them with ball ammunition.

■ **49. EXAMINATION.**—*a.* Prior to the date of actually firing the submachine gun for record, each gunner so firing is required to pass the examination of preparatory training shown on the form in paragraph 102. The date of this examination is recorded on each individual's qualification record card.

   b. The questions given herein are examples for the examination. Men are required to explain them in their own words or demonstrate them by their own actions. These questions are given only as a guide. Any pertinent question under the subjects listed on the form in paragraph 102 should be asked.

Q. Name the parts of the weapon as I point to them.—*A.* Pupil names each part as pointed out.

Q. How many rounds of ammunition can be placed in the magazines?—*A.* The box magazine holds 20 rounds; the drum magazine holds 50 rounds.

Q. Remove the frame group and show me the extractor.—*A.* Pupil removes frame group and shows the instructor the extractor.

Q. What is the first thing to do to assemble the trigger mechanism?—*A.* See that the magazine catch is in the assembled position.

Q. When should the weapon be loaded, and what should be done before loading?—*A.* Only when you are ready to fire. The safety lever should be set to "safe" before loading.

Q. Should the bolt be closed with a loaded magazine inserted? Why?—*A.* No, because the forward movement of the bolt loads a round into the chamber and fires the weapon.

Q. In case of misfire what should first be done?—*A.* The actuator knob should be pulled to the rear to insure ejection of a misfired cartridge, and the safety should then be turned on.

Q. Demonstrate the standing, kneeling, or prone position.—*A.* Pupil demonstrates the required position.

Q. Where is the battle sight?—*A.* Pupil points out the battle sight.

Q. How do you set a range setting of 200 yards and a right windage of 1 point?—*A.* Pupil demonstrates setting.

Q. Does the rear sight remain stationary in firing?—*A.* No, it moves forward with the bolt when the trigger is pulled.

Q. Demonstrate how to load a drum magazine.—*A.* Pupil demonstrates method of loading drum magazine.

Q. What parts of the gun should be cleaned after firing each day?—*A.* The bore and chamber, and all parts and surfaces of the receiver, bolt, ejector, and extractor that are contacted by the powder gases.

SECTION II

COURSES TO BE FIRED

■ 50. ARMY REGULATIONS APPLICABLE.—AR 775–10 prescribes details as to who will fire and ammunition allowances.

■ **51. INSTRUCTION PRACTICE.**—*a.* The following table prescribes the firing in instruction practice in the order followed by the individual soldier. The table is fired three times for instruction.

QUICK FIRE—TARGETS E, F, AND M

Phase	Type of fire	Position	Range	Time	Shots
A____	Single shot_____	Standing____	15–35 yards__	Each target exposed 3 seconds.	10 shots, 2 per target.
B____	Automatic bursts of 3.	Standing or kneeling.	25–30 yards__	Each group exposed 5 seconds.	15 shots, 3 per target.

b. There is a 5-second interval after the completion of phase A, at which time the timer blows a whistle and the firer starts walking from position A to position B. Phase B starts 10 seconds after the whistle.

■ **52. RECORD PRACTICE.**—The table used for instruction practice is fired once for record.

SECTION III

CONDUCT OF RANGE PRACTICE

■ **53. GENERAL.**—Organization commanders are responsible for conducting the range practice of their organizations in accordance with the provisions of this manual and applicable Army Regulations. All firing is done under the direct supervision of a commissioned officer. No person is permitted to start range practice until after he has successfully passed the gunner's examination as shown in preparatory training.

■ **54. PROCEDURE** (fig. 15).—*a. Firing phases.*—The gunner takes his place at the initial position (fig. 15) with a drum loaded with 25 rounds of ammunition.

(1) *Phase A.*—At the order of the officer in charge of firing, the gunner loads his piece, sets it for semiautomatic

41

fire, and calls "Ready." The officer in charge then blows his whistle and 3 seconds thereafter the targets in group A are exposed in irregular order, one at a time, for 3 seconds each. Three seconds after the first target disappears,

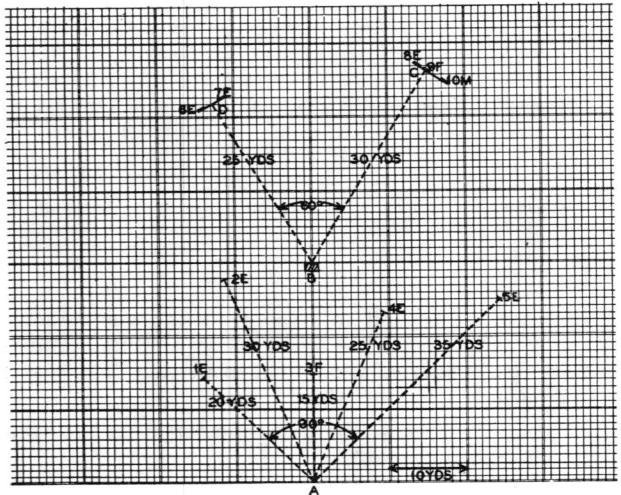

LEGEND:
A – INITIAL POSITION OF GUNNER (STANDING).
B – PIT
C – 3 TARGETS OPERATE AS ONE
    1 YD BETWEEN KNEELING AND PRONE SILHOUETTES, CENTER TO CENTER
    2 YDS BETWEEN PRONE AND STANDING SILHOUETTES, CENTER TO CENTER
D – 2 TARGETS OPERATE AS ONE
    3 YDS BETWEEN SILHOUETTES CENTER TO CENTER
PRONE SILHOUETTE TARGET F
KNEELING SILHOUETTE TARGET E
STANDING SILHOUETTE TARGET M

ALL TEN TARGETS ARE BOBBING TARGETS OPERATED BY ROPES.

FIGURE 15.—Dismounted practice course.

another appears, and so on. The targets are controlled from in rear of the gunner as explained in section IV. The gunner is allowed two shots per target in this phase. The standing position is used. The gunner may keep the gun to his shoulder at all times. All firing ceases 60 seconds after the first whistle signal.

42

(2) *Phase B.*—(a) As soon as the fifth target disappears, the gunner sets his gun for full automatic fire. Five seconds after the fifth target disappears, the timer blows his whistle. At this signal the gunner starts walking toward the pit (B). Ten seconds after the whistle signal, target group C (or D) is exposed for 5 seconds, during which time the gunner fires a burst of 3 shots at each target of the group. Five seconds after the target group C (or D) disappears, the remaining group D (or C) is exposed for 5 seconds, during which time the gunner fires his remaining ammunition.

(b) Between exposures of group C (D) and group D (C) the gunner may remain halted or move, but he may not advance beyond a barrier just short of the pit (B).

(c) In phase B the gunner may fire standing or kneeling.

(d) All firing ceases 60 seconds after the first whistle signal.

b. Action on completion of course.—Upon completion of the course, the gunner removes the drum, unloads, and reports that his gun is clear. This fact is verified by a noncommissioned officer before anyone is permitted to move in front of the gun.

c. Timer.—(1) A commissioned or noncommissioned officer takes the time for each run. He is responsible for the second whistle signal described in a (2) (a) above.

(2) Two assistant timers, one in rear of position A (fig. 15) and one in the pit (B, fig. 15), record the intervals (3 and 5 seconds, respectively) during which the targets are exposed and withdrawn.

d. Coaches.—(1) *Instruction practice.*—Considerable time and effort can be saved if well-qualified coaches are used at firing points in instruction practice to instruct the gunners. In view of the fact that all fire is quick fire and the run is not interrupted once it is started, best results are obtained by having the coach observe the actions of the gunner during each run and then, after the targets are marked, having both gunner and coach report to a designated place, where the coach points out errors and gives the necessary instruction to prevent their repetition on the next run. Prior to, during, and after the run, coaches require gunners to employ habitually all practical safety measures.

(2) *Record practice.*—No coaching is permitted during record practice.

e. Penalties.—For any infraction of the rules of procedure laid down herein, the gunner is penalized 5 points for each round fired improperly.

f. Possible score.—100 points.

g. Form for score card.—See paragraph 102.

h. Classification.—The individual classification to be attained and the minimum aggregate scores required for qualification are as prescribed in AR 775–10.

■ 55. RANGE FIRING.—*a. Marking, scoring, etc.*—(1) The targets are designated 1, 2, 3, 4, and 5 from left to right in group A; 6 and 7 in group D; and 8, 9, and 10 in group C. (See fig. 15.)   The number of men detailed as markers is left to the discretion of the officer conducting the firing; normally they will be the men who pull the targets in phase A. The markers take position in rear of position A (fig. 15) and await the completion of phase B.   After phase B is completed, they run to their designated targets, examine them, and face the scorer.   One noncommissioned officer detailed as scorer waits at a convenient place.   The markers then call out in numerical order the hits or misses, for example, No. 1, a hit or 2 hits; No. 2, a miss, etc.   After calling the hits and misses in group A, the markers cover any shot holes with pasters.   Two markers then run to groups D and C, respectively, mark the shots, call, and paste them.

(2) During record practice, scores are recorded by a noncommissioned officer of a different organization, if practicable, under the supervision of an officer who personally checks the score card, authenticates it, and retains all cards in his personal possession, except while the man is actually firing.

b. Value of hits.—Each target hit counts 5 points, and each hit on any target (not to exceed 2 for phase A and 3 for phase B) counts 2 points.

c. Defective cartridges and malfunctions.—If a defective cartridge or a malfunction causes failure to fire at any one of the targets, the firer locks his piece and indicates this fact to the officer in charge of firing, who blows his whistle and time is stopped.   The malfunction is reduced or a new cartridge is inserted in the chamber. When ready, he signals or otherwise informs the officer in charge. At the second whistle, time is in again and the phase is continued.

SECTION IV

TARGETS, RANGES, AND RANGE PRECAUTIONS

■ 56. TARGETS (fig. 16).—*a.* Target F is a drab silhouette representing a prone figure.

b. Target E is a drab silhouette representing a kneeling figure.

c. Target M is a drab silhouette representing a standing figure. It is in two parts, the upper is target E and the lower is a trapezoidal piece whose upper edge is placed closely against the lower edge of target E.

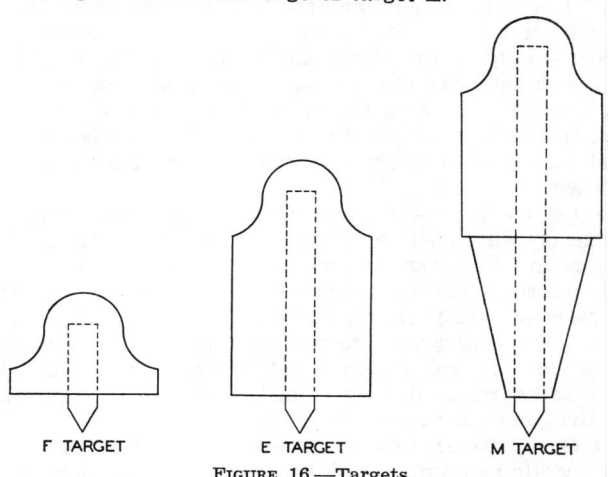

F TARGET   E TARGET   M TARGET

FIGURE 16.—Targets.

■ 57. CONSTRUCTION OF THE COURSE.—*a. Location.*—The course may be laid out on any ordinary terrain, preferably wooded, with some grass, weeds, and underbrush. Targets should be partially concealed or be near possible concealment in order to represent as nearly as practicable actual enemy groups.

b. Lay-out.—(1) The course is laid out as shown in figure 15. Distances and targets are as indicated in the figure. There are two groups of targets, each necessary for one phase of the firing. Group A is a group of five targets, 4 kneeling (E) and 1 prone (F), which are engaged from position A. Group B consists of two groups (D and C, fig. 15)—one to the left front consisting of two kneeling targets (E); and one to the right front consisting of one standing target (M), one kneeling (E), and one prone (F). In group B, the whole group (C or D) operates as a unit in appearing and disappearing.

(2) The pit consists of an emplacement of concrete (or other suitable material), proof against .45 caliber ammunition, with suitable rear and overhead cover but with an opening in front. It is about 3 feet by 6 feet, with the floor at least 3 feet in the ground and the roof not over 1 foot above the level of the ground. It should hold two men and the necessary equipment to control the targets of groups C and D.

c. Operation.—(1) *General.*—All targets are bobbing targets, i. e., they are so arranged that they can be fully exposed to the firer, or turned so that only the edge of the target points toward the firer during the time that the target is not exposed.

(2) (*a*) *Group A.*—Each target in this group must be capable of being operated individually from in rear of the firing point. This can be done with two ropes to a target, or by one rope only if a spring arrangement is used to hold the target with only the edge exposed. The use of one rope, which is run through a pipe, to each target will prevent the gunner from knowing which target is to be exposed next. If visible ropes are used, they must be kept tight during the time the gunner is at the firing point.

(*b*) *Group B.*—In this group, the targets at D and at C each operate as a unit. For example, when the group at D is operated, both the E targets are exposed and withdrawn together. The targets are operated from the pit (B, fig. 15). Other details are as in (*a*) above.

(3) *Method of signaling.*—As the officer in charge, the timer, and the target operators are in rear of the gunner (in phase A), the officer can signal the target he desires to be

exposed next by the number of fingers he exposes. (See par. 55 for numbering of targets.)

■ 58. RANGE PRECAUTIONS.—(See also par. 20 and 38.) During firing, all personnel including marking details must be in safe positions. The necessary range guards are posted and danger flags prominently displayed before firing begins. There must be no firing until so ordered by the officer in charge. The provisions of AR 750–10 must be complied with during all range firing.

CHAPTER 4

MARKSMANSHIP, MOVING GROUND AND AIR TARGETS

SECTION I

MOVING GROUND TARGETS

■ 59. GENERAL.—All units armed with the submachine gun will be trained to fire at moving targets, both vehicular and personnel. Normally such fire will be delivered at short ranges in short bursts of fire. The high rate of fire and ability of the submachine gunner to move the trajectory of fire at will make the submachine gun particularly effective against moving personnel, either individual or in groups. Training of the submachine gunner must be such as to enable him to employ his gun effectively and quickly. To this end he must be trained in the proper use of sights and methods of leading at short ranges.

■ 60. SIGHTS.—Moving targets are seldom exposed for long periods and can be expected to move at maximum speed during periods of exposure. Moving personnel are especially difficult to hit. Since the submachine gun is essentially a short range weapon of opportunity, the battle sight is habitually used in firing at moving targets. No windage adjustment it attempted. When necessary to fire near the maximum effective range (about 350 yards) or at greater range, and time is available, the rear sight scale may be set at the proper range.

■ 61. LEADS.—Targets that cross the line of sight require the gunner to aim ahead of the target so that the paths of the target and bullet will meet. The distance ahead of the target is called the "lead." Targets which approach directly

48

toward the gunner or recede directly from the gunner require no lead. For personnel targets moving across the line of sight, the point of aim should be slightly in front of the body and the lead corrected by observation of the effect of the spray of the rounds fired.

■ 62. DETERMINATION OF LEADS.—The lead necessary to hit a moving vehicle is dependent upon the speed of the target, the range to the target, and the direction of movement with respect to the line of sight. Moving at 10 miles an hour a vehicle moves approximately its own length of 5 yards in 1 second. The velocity of a bullet from the submachine gun is approximately 800 feet or slightly more than 250 yards in 1 second. Therefore to hit a vehicle moving at 10 miles an hour at ranges of about 250 yards and 500 yards, the leads should be 5 yards and 10 yards, respectively. At a speed of 20 miles an hour the leads should be 10 yards and 20 yards, respectively, etc.

■ 63. APPLICATION OF LEADS.—a. Leads are applied by using the length of the target as it appears to the gunner as the unit of measure. This eliminates the necessity for corrections due to the angle at which the target crosses the line of sight, because the more acute the angle the smaller the target appears and the less lateral speed it attains.

b. The following lead table for vehicles is furnished as a guide:

Miles per hour	Range		
	125 yards or less	250 yards	500 yards
10	½ TL	1 TL	2 TL
20	1 TL	2 TL	4 TL

■ 64. TECHNIQUE OF FIRE.—a. The following technique is used by the gunner for firing at moving targets:

(1) *Approaching or receding targets.*—The gunner holds his aim on the center of the target and fires automatic in short bursts.

49

(2) *Crossing vehicular targets.*—The gunner estimates the proper number of leads, alines his sights on the bottom of the target at its rearward point, swings straight across the target to the estimated lead, and fires short bursts of fire keeping the weapon at the proper lead.

(3) *Crossing personnel targets.*—The gunner takes aim slightly in front of the center of the body of the target and fires short bursts. The lead is changed an appropriate amount after observation of the effect of the spray of bursts.

b. The high rate of fire of the submachine gun allows the gunner to spray the target with fire and improve his lead estimation by actual observation of the effectiveness of his fire.

■ 65. PLACE IN TRAINING.—Firing at moving targets with service ammunition should follow instruction in known distance firing and firing of the dismounted practice course.

SECTION II

AIR TARGETS

■ 66. GENERAL.—Combat arms take the necessary measures for their own immediate protection against low flying hostile aircraft. All available weapons are normally employed in this defense. Consequently all units armed with the submachine gun will be trained in the use of this weapon against air targets. The low muzzle velocity, short effective range, and the tactical employment of the submachine gun are factors which mitigate against its effective use on air targets. Normally the submachine gun is issued as a supplemental weapon to vehicle crews, and more effective weapons, such as the caliber .30 and caliber .50 machine guns, are available and will be employed for immediate fire at aircraft. In comparison with these other weapons, submachine gun fire is relatively ineffective against hostile aircraft. Proper training in the use of sights and methods of leading will increase the effectiveness of fire and are essential phases of the training of the submachine gunner. Application of the fundamentals of firing against moving ground targets are satisfactory for training the submachine gunner in firing against air targets. Special use of sights and leads for air targets are included in paragraphs 67 to 70, inclusive.

■ 67. SIGHTS.—The battle sight is habitually used for firing on low flying airplanes. Hostile airplanes move so rapidly that time is not available for setting of sights.

■ 68. LEADS.—In order to hit an air target in flight, it is necessary to aim an appropriate distance ahead of it and on its projected path of flight so that the target and the bullet will meet. This distance ahead of the airplane is called LEAD. A lead must be applied to all firing, except when the target is at an extremely close range (100 feet), when it is diving directly at the gunner, or flying directly from him.

■ 69. DETERMINATION OF LEADS.—*a.* The lead necessary to engage any target depends upon the—
  (1) Speed of the target.
  (2) Range to the target.
  (3) Time of flight of the bullet.
  (4) Direction of flight of the target with respect to the line of fire.

  b. When a target appears, it is impossible for submachine gunners to consider all the factors listed above and to compute accurately the lead required for firing. Use of a quick rule of thumb and experience and proficiency in firing are essential.

  c. Based on an average plane of 30 feet in length moving at 200 miles an hour, the plane moves 10 times its own length in 1 second. The velocity of a bullet from the submachine gun is approximately 800 feet or slightly more than 250 yards in 1 second. Therefore to hit an airplane moving across the front at 200 miles an hour at a height of 800 feet, the aiming point of the gunner is 10 leads in front of the airplane, and at a height of 400 feet the aiming point is 20 leads in front of the airplane.

■ 70. APPLICATION OF LEADS.—*a.* Since it is impracticable for the submachine gunner to estimate accurately the speeds and heights of attacking airplanes, a general rule is given for firing at air targets. Application of this rule by the

51

individual, coupled with the proper fire distribution of other weapons, will form a mass of fire of which some portion will be effective against the air target.

b. The following lead table for air targets is given as a guide:

Miles per hour	Height of plane		
	100 to 600 feet	600–1,000 feet	Over 1,000 feet
200	20 TL	10 T L	.5 T L

c. Leads are applied by using the length of the target as it appears to the gunner as the unit of measure. The ability of the submachine gunner to spray his bursts of automatic fire constitutes a balance against the errors in leading by the above general rule.

■ 71. TECHNIQUE OF FIRE.—The following technique should be used by the gunner in firing at air targets:

a. For direct-diving airplanes, direct-climbing airplanes, or airplanes at less than 100 feet height, the gunner holds his aim on the center of the airplane and fires bursts of automatic fire.

b. For all other airplanes, the gunner aims at the airplane, swings to the proper number of leads along the path of the airplane, and fires short bursts of fire, keeping the weapon at the proper lead.

■ 72. PLACE IN TRAINING.—Firing at air targets with service ammunition is the last phase of instruction for the submachine gunner and follows instruction in vehicular firing described in chapter 5.

SECTION III

MOVING TARGETS, RANGES, AND RANGE PRECAUTIONS

■ 73. GENERAL.—The ability of the submachine gunner to hit a target moving on the ground or in the air is developed through appropriate exercises conducted as part of the combat firing of his organization. The following courses are included as exercises to achieve this result. In the moving ground target course, the gunner fires at a moving ground target from a vehicle that moves between bursts but is halted when the gunner actually fires. In the air target course, the gunner fires from a stationary vehicle. Firing at towed air targets follows this course.

■ 74. TARGETS.—The target consists of a rectangular frame 5 feet by 8 feet in size, longer axis horizontal, covered with target cloth or other light-colored material mounted on a suitable carriage which has the ability to move at a maximum speed of 20 miles per hour. The substitution of E and M targets for the target cloth provides a suitable target for training in firing at moving personnel. The addition of a superstructure at the desired height holding a small air target provides a suitable target for training in firing at air targets.

■ 75. RANGE CONSTRUCTION.—*a.* The moving target range should be constructed generally as shown in figures 17 and 18.

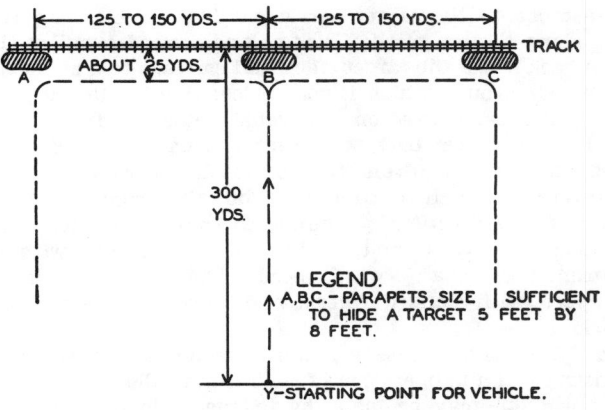

FIGURE 17.—Moving ground target.

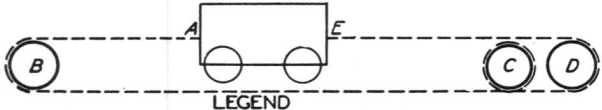

LEGEND

DOTTED LINE REPRESENTS A STEEL CABLE
B AND D — PULLEYS
C — CYLINDRICAL DRUM ATTACHED TO MOTOR

FIGURE 18.—Moving target set-up.

b. *A, B,* and *C* represent parapets of sufficient size to hide completely the target from the gunner's view, with dugouts in the rear for sheltering pit details. These parapets should be 125 to 150 yards apart. The entire distance from *A* to *C* should be equipped with a suitable continuous track behind the parapets for movement of the target. *Y* is the starting point of the gunner's vehicle, and *YB* the direction in which the gunner's vehicle moves. The track for the vehicle should be smooth and level. The entire area from *Y* to the parapets should be clear in order that ground firing points may be employed for instruction in firing at moving personnel and air targets.

c. The dotted line (fig. 18) represents a steel cable which is fastened to the target carriage. The cable is run from (A) along the top of the track to and around a pulley (B); then back under the target track and carriage to a cylindrical drum (C) around which it goes twice; then around a pulley (D) which is mounted on a movable frame to adjust tension in the cable; then back to the target carriage at (E). The shaft (or axle) of drum (C) is attached through a transmission and clutch to a motor. Thus the drum may be rotated in one direction by running the motor with the transmission in reverse and in the other direction with the transmission in a forward speed. The target carriage is equipped with flanged wheels and is run on narrow gage tracks.

d. When the necessary material is not available for the construction and operation of such an installation, a simpler arrangement may be made by towing a double-ended sled behind a vehicle; but with a sled, the target cannot be so readily stopped and started again in either direction.

■ 76. OPERATION OF RANGE AND COURSE TO BE FIRED FOR MOV-
ING GROUND TARGETS.—*a.* (1) *First run.*—The gunner starts
at *Y* mounted in the vehicle with his weapon loaded and
locked on automatic; the vehicle moves at 15 to 20 miles per
hour toward *B*. The target may be behind either parapet
A or *C*. Upon telephone or visual signal from behind the
starting line, the target is released and moved at 10 to 20
miles an hour to parapet *B*. When the target appears, the
gunner's vehicle is stopped, the gunner unlocks his weapon
and fires from the stationary vehicle at the moving target as
long as it is visible. The target should be released so that
the gunner fires the first run at a range of not more than 250
or less than 150 yards.

    Rounds fired: 20.

    Method of fire: Bursts of about 3 rounds fired in rapid
               succession.

  (2) *Second run.*—The gunner quickly reloads with a new
clip of 20 rounds, locks the piece, and calls "Ready," where-
upon the vehicle again moves forward at 10 to 20 miles per
hour. The target should be held behind parapet *B* after the
first run for 30 seconds, at the end of which time it is again
released in either direction. During the second run, the
gunner again fires from a stationary vehicle as before as
long as the target is visible.

    Rounds fired: 20.

    Method of fire: Optional.

  b. Scoring.—10 points for hitting the target; 2 points for
each hit on the target. Maximum score: 100.

■ 77. OPERATION OF RANGE AND COURSE TO BE FIRED FOR AIR
TARGETS.—*a.* (1) *First run.*—The gunner stands at a ground
firing point immediately in front of parapet *B* with the submachine gun loaded and locked on automatic. The moving
target modified with a small superstructure of the desired
height is placed behind parapet *B*. Upon signal by the officer
in charge of firing, the target is released and moves toward
either parapet *A* or *C*. Five seconds after the target has
begun to move, the gunner may commence firing and may
continue to fire until his ammunition is exhausted.

    Rounds fired: 20.

    Method of fire: Optional.

(2) *Second run.*—The gunner stands at a ground firing point immediately in front of parapet *B* with the submachine gun loaded and locked on automatic. The moving target, modified with a small superstructure of the desired height, is placed behind parapet *A* or *C*. Upon signal by the officer in charge of firing, the target is released and moves toward parapet *B*. Five seconds after the target has begun to move, the gunner may commence firing and may continue to fire until his ammunition is exhausted.

Rounds fired: 20.

Method of fire: Optional.

b. Scoring.—10 points for hitting the target on each run; 2 points for each hit on the target. Maximum score: 100.

■ 78. RANGE PRECAUTIONS.—(See pars. 20, 38, and 58.) Firing takes place only at such times as a flag is properly displayed by the target detail indicating that they have taken adequate cover. In instruction firing at air targets and firing at towed air targets, the right and left limits of fire will be plainly marked by posts or flags.

CHAPTER 5

VEHICULAR FIRING

Section I

GENERAL

■ **79. GENERAL.**—Units armed with the submachine gun are trained to fire from stationary and moving vehicles at appropriate targets. Practice in firing from moving vehicles follows instruction in known distance firing, dismounted. Two instruction courses are described in this chapter, one for firing from open vehicles and one for firing from turreted vehicles. For vehicular firing, vehicles are closed as for combat except that ports may be open. Speeds are as nearly uniform as practicable and generally conform to speeds indicated in each course. The vehicle is equipped with the full number of weapons authorized, all in normal position for combat.

Section II

OPEN VEHICLE

■ **80. ARMY REGULATIONS APPLICABLE.**—AR 775–10 prescribes details as to who will fire and ammunition allowances.

■ **81. INSTRUCTION PRACTICE.**—The following table prescribes the firing in instruction practice in the order followed by the individual soldier. This practice is fired twice. One 50-round drum is allowed to fire the table.

57

QUICK FIRE—TARGETS E AND F

Phase	Type of fire	Range	Vehicular speed	Shots
A	Automatic_____	75–25 yards____	20 miles per hour_____	15
B	Automatic_____	65–5 yards_____	20 miles per hour_____	10
C	Semiautomatic_____	12–75 yards____	Halted and 20 miles per hour_	25

■ 82. Record Practice.—Vehicular firing is not included in firing in record practice.

■ 83. Procedure (fig. 19).—*a. Firing phases.*—(1) The gunner takes his place in the front compartment of the vehicle on the track about 50 yards from point *A* with a loaded 50-round drum.

(a) *Phase A.*—At the order of the officer in charge of firing, the gunner loads his piece, sets it for automatic fire, and calls "Ready." The officer in charge directs the vehicle to start on the course. The vehicle gains and maintains a uniform speed of approximately 20 miles per hour between halts. The gunner may start firing on the targets of group 1 as soon as the vehicle passes point *A*, at which point the timer starts taking time for the run. The gunner remains low in the vehicle and exposes himself only enough to fire about fifteen rounds at the targets in group 1. He ceases firing on these targets upon reaching point *B* which is 75 yards from point *A*.

(b) *Phase B.*—The vehicle continues on the course, and after making the turn and reaching the point *E* the gunner may commence firing on targets of group 2, all to his right front. He fires about ten rounds at targets of this group. He ceases firing when his vehicle passes the point *G*.

(c) *Phase C.*—After passing group 2, the gunner moves to the rear of the vehicle, sets his weapon on semiautomatic, and prepares to engage targets of group 3. The vehicle continues on the course until it reaches point *J* where it halts for 15 seconds. The gunner can commence firing on

58

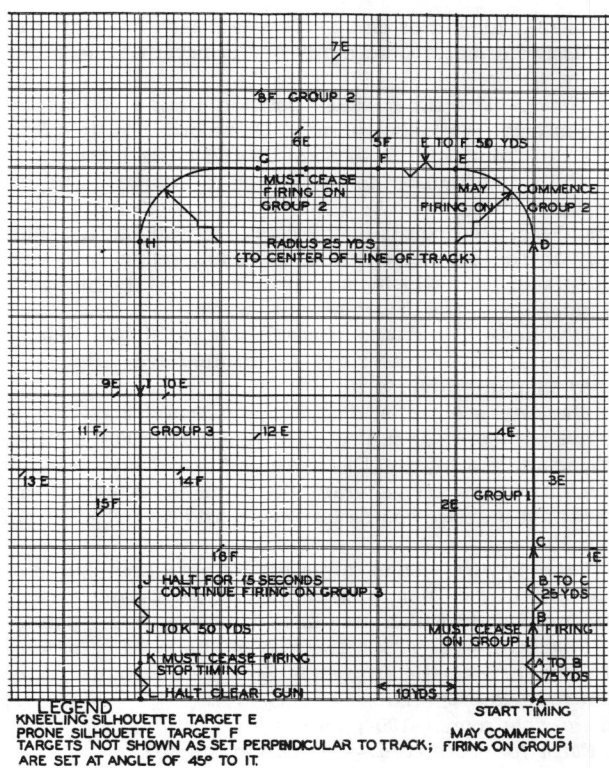

FIGURE 19.—Course for firing from open vehicles.

targets of group 3 as soon as the vehicle comes to a full halt. He may continue to fire after the vehicle resumes moving on the course, but ceases firing upon expenditure of the 50-round drum or when the vehicle passes point *K*.

(2) All firing in all phases must cease in the time necessary to travel the distance from point *A* to point *K* at 20 miles per hour, plus 25 seconds.

b. Action on completion of course.—Upon completion of the course at point *L*, the gunner removes the drum, unloads, and reports that his gun is clear. This fact is verified by a non-commissioned officer before any one is permitted to move in front of the gun.

c. *Timer.*—-The timer takes position in the vehicle so as not to interfere with the gunner. He records the time for the entire run from point *A* to point *K* and reports the time for each individual gunner to the officer in charge. The timer may assist the gunner in commencing and stopping his fire at the designated points in the course by tapping the gunner lightly on the back or shoulder or commanding CEASE FIRING at the proper points during the run. He times the halt periods and indicates to the driver when to move.

d. *Coaches.*—The duties of coaches for vehicular firing are the same as those prescribed for dismounted firing as described in paragraph 54 *d.*

e. *Marking, scoring, and defective ammunition.*—The targets are numbered as indicated in figure 19. The method of marking and scoring and the treatment of defective cartridges and malfunctions are generally the same as prescribed in the dismounted practice course for record and described in paragraph 55.

f. *Value of hits.*—Each target hit counts 5 points, and each hit on any target (not to exceed 3 hits per target, except on target number 16 which may have 5 hits) counts 2 points.

g. *Penalties.*—For any infraction of the rules of precedure laid down herein, the gunner is penalized 5 points for each round fired improperly.

h. *Possible score.*—180 points.

i. *Form for score card.*—See paragraph 102.

■ 84. CONSTRUCTION OF THE COURSE.—*a. Location.*—The course may be located on any ordinary terrain with some grass, weeds, and underbrush and a good field of fire for the gunner. Targets should be partially concealed but at the same time they must be easily located by the gunner. The track should be constructed so as to be reasonably smooth. The vehicle should travel in the center of the track.

b. *Lay-out.*—The course is laid out as shown in figure 19. Distances are as shown and targets are as indicated in the legend. There are three groups of targets, each necessary for one phase of the firing. Group 1 is a group of 4 targets

(4 kneeling, target E); group 2 is a group of 4 targets (2 kneeling, target E, and 2 prone, target F); group 3 is a group of 8 targets (4 kneeling, target E, and 4 prone, target F). All targets are fixed targets.

Section III

TURRETED VEHICLE

■ 85. Army Regulations Applicable.—AR 775–10 prescribes details as to who will fire and ammunition allowances.

■ 86. Instruction practice.—The following table prescribes the firing in instruction practice in the order followed by the individual soldier. This practice is fired twice. One 50-round drum is allowed to fire the table.

QUICK FIRE—TARGETS E, F, AND M

Phase	Type of fire	Range	Vehicular speed	Shots
A	Automatic_____	90–10 yards____	20 miles per hour_____	20
B	Semiautomatic_____	10–40 yards____	Halted and 20 miles per hour.	20
C	Semiautomatic_____	15–70 yards____	20 miles per hour_____	10

■ 87. Record Practice.—Vehicular firing is not included in firing in record practice.

■ 88. Procedure (fig. 20).—*a. Firing phases.*—(1) The gunner takes his place in the vehicle on the track about 50 yards from point *A* with a loaded 50-round drum. All submachine gun firing is through pistol ports.

(*a*) *Phase A.*—At the order of the officer in charge of firing, the gunner loads his piece, sets it for automatic fire, and calls "Ready." The officer in charge directs the vehicle to start on the course. The vehicle gains and maintains a uniform speed of approximately 20 miles per hour between halts. The gunner may start firing on the targets of group 1

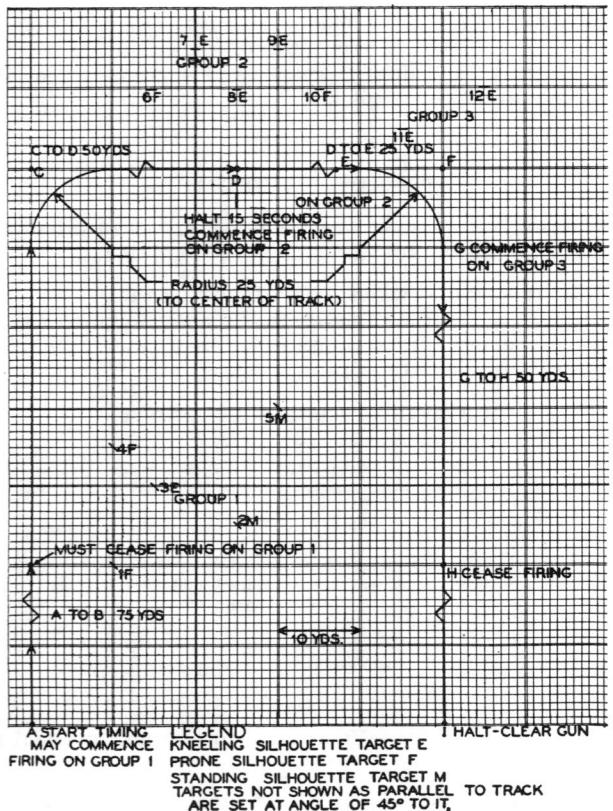

FIGURE 20.—Course for firing from turret-top vehicles.

as soon as the vehicle passes point *A,* at which time the timer starts taking time for the run. The gunner fires about twenty rounds between points *A* and *B,* using short bursts of automatic fire on each target. At point *B* he ceases firing on targets of group 1.

(b) *Phase B.*—As the vehicle moves toward point *D* the gunner sets the gun on semiautomatic and prepares to fire from the left and rear ports. Upon reaching point *D* the vehicle halts for 15 seconds. When the vehicle has come to a full halt, the gunner may commence firing on the targets of group 2. The vehicle moves forward after the 15-second halt. The gunner may continue to fire after the vehicle has moved from point *D*, but must cease firing when point *E* is reached.

(c) *Phase C.*—The vehicle continues to move toward point *G*, and the gunner prepares to fire semiautomatic fire from the rear ports on targets of group 3. He may commence firing on these targets when the vehicle passes point *G*, but must cease firing when the vehicle reaches point *H*.

(2) All firing in all phases must cease in the time necessary to travel the distance from point *A* to point *H* at 20 miles per hour, plus 25 seconds.

b. Action upon completion of course.—Upon completion of the course at point *I*, the gunner removes the drum, unloads and reports that his gun is clear. This fact is verified by a noncommissioned officer before anyone is permitted to move in front of the gun.

c. Timer.—The timer takes position in the vehicle so as not to interfere with the gunner. He records the time for the entire run from point *A* to point *H* and reports the time for each individual gunner to the officer in charge. The timer may assist the gunner in commencing and stopping his fire at the designated points in the course by tapping the gunner lightly on the back or shoulder or commanding CEASE FIRING at the proper points during the run. He times the halt periods and indicates to the driver when to move.

d. Coaches.—The duties of coaches for vehicular firing are the same as those prescribed for dismounted firing as described in paragraph 54 *d.*

e. Marking, scoring, and defective ammunition.—The targets are numbered as indicated in figure 20. The method of marking and scoring and the treatment of defective cartridges and malfunctions are generally the same as prescribed in the dismounted practice course for record and as described in paragraph 55.

63

f. Value of hits.—Each target hit counts 5 points, and each hit on any target (not to exceed 4 hits per target, except on target number 12 which may have 6 hits) counts 2 points.

g. Penalties.—For any infraction of the rules of procedure laid down herein, the gunner is penalized 5 points for each round fired improperly.

h. Possible score.—160 points.

i. Form for score card.—See paragraph 102.

■ 89. CONSTRUCTION OF THE COURSE.—*a. Location.*—The course is located and laid out in the same manner as described in paragraph 84 for vehicular firing from open top vehicles, but conforming to figure 20.

b. Lay-out.—The course is laid out as shown in figure 20. Distances are shown and targets are as indicated in the figure. There are three groups of targets, each necessary for one phase of the firing. Group 1 is a group of 5 targets (1 kneeling, target E; 2 prone, target F; and 2 standing, target M); group 2 is a group of 5 targets (3 kneeling, target E, and 2 prone, target F); group 3 is a group of 2 targets (2 kneeling, target E). All targets are fixed targets.

CHAPTER 6

TECHNIQUE OF FIRE

■ 90. CHARACTERISTICS OF FIRE.—*a. General.*—The characteristics of the submachine gun govern the manner in which it is to be used. It is essentially an individual weapon like the rifle, pistol, or bayonet; but it does not supplant or replace any of the others. It is highly effective at close quarters, because it is light, compact, and, with a 50-round drum, can remain in action when other weapons have expended their ammunition. The gun is very dependable due to the simplicity of its mechanism. Efficient gunners can be quickly developed. The gunner has 20 to 50 rounds immediately available, and by changing magazines he can pour a considerable volume of fire into an enemy that confronts him. This volume of fire, and the flexibility of fire of which the individual gunner is capable, are important characteristics of the submachine gun, because of the ease of reloading and shifting the fire when fighting at close quarters. The fact that the weapon can be fired either as a single-shot gun or with full automatic fire is also important. In terrain where cover and concealment are abundant, especially against riflemen or machine gunners, this flexibility of fire is an advantage. The gun is equipped with a compensator which takes up much of the recoil and permits the gunner who uses the weapon properly to attain considerable accuracy of fire even when full automatic fire is used.

b. Comparison with other weapons.—As compared to the M1903 or M1 caliber .30 rifle, the caliber .45 submachine gun is at a disadvantage in that it has less killing power, it is less accurate at ranges over 150 yards, and has no bayonet. However, the submachine gun has the distinct advantage of being well suited to combat in close quarters.

c. Collective firing.—Collective firing is the combined firing of a group of individuals. The submachine gun is normally issued as a vehicular or individual weapon. It is not issued to all members of a unit as is the pistol or the rifle. Consequently, collective firing of submachine guns is not employed. The submachine gun may be used in conjunction with other weapons, especially the machine gun. When used in this manner it is normally employed at short-range targets, while the alternate weapon fires at relatively long range. The weapon may be used by motorcycle scouts operating individually at extremely short ranges.

■ 91. TYPE OF FIRE ORDERS.—*a.* Formal fire orders are seldom necessary or desirable. For control on the target range such orders as INSERT LOADED MAGAZINE, READY, COMMENCE FIRING, and CEASE FIRING are used. In combat, fire orders if necessary at all are normally limited to COMMENCE FIRING, and CEASE FIRING.

b. The target designation may be added to the fire order when the target has not been discovered by the submachine gunner. In this case, the fire order may be given to the submachine gunner as follows, "Jones, rifleman behind tree to right front, commence firing."

■ 92. TARGET DESIGNATION.—The normal employment of the submachine gun is such as to preclude the necessity for the formal target designation desirable for longer-range weapons employed in groups. Targets which are not perfectly obvious may be designated by pointing, by oral description, or by firing upon them with tracer ammunition. Normally targets are discovered and immediately taken under fire by the individual gunner himself.

■ 93. RANGE ESTIMATION.—*a. General.*—In battle, ranges are seldom known in advance, so that the effectiveness of fire depends in large measure upon the accuracy of range estimation. The gunner is trained in the estimation of ranges up to about 500 yards for proper sight setting. He is also made proficient in determining the correct point of aim with the battle right at all ranges up to about 200 yards.

b. Appearance of objects.—In some cases much of the ground between the observer and the targets is hidden from view. In such cases the range is estimated by the appearance of objects. Whenever the appearance of objects is used as a basis for range estimation, the observer must make allowance for the following effects:

(1) Objects seem nearer—

(a) When the object is in a bright light.

(b) When the color of the object contrasts sharply with the color of the background.

(c) When looking over water, snow, or a uniform surface like a wheat field.

(d) When looking downward from a height.

(e) In the clear atmosphere of high altitudes.

(f) When looking over a depression, most of which is hidden.

(2) Objects seem more distant—

(a) When looking over a depression most of which is visible.

(b) When there is a poor light or fog.

(c) When only a small part of the object can be seen.

(d) When looking from low ground upward toward higher ground.

c. *Training in range estimation.*—Proficiency in range estimation can be obtained only by constant practice and diligence. The methods and exercises prescribed in paragraph 46 are suitable means of developing proficiency in range estimation.

■ 94. SAMPLE PROBLEMS.—*a. General.*—In preparing exercises involving the use of the submachine gun, advantage is taken of field exercises and maneuvers to present logical situations, some phases of which would require the employment of this weapon, both from the ground and from the vehicle. These exercises should include the use of the submachine gun in the dismounted reconnaissance of a road block, employment on outpost duty or in establishing march outpost, and use by motorcycle scouts and by armored vehicle personnel in assumed ambush situations.

b. Exercises.—The following exercises are given as a guide and may be modified to suit the terrain, equipment, and time available. Each problem utilizes natural terrain features, equipment normally available, and actual personnel targets. The exercise should be conducted under the supervision of

67

a commissioned officer who will make all checks and point out all errors. Service ammunition is not fired during these exercises. The purpose of all exercises is to train the individual submachine gunner and unit leaders in target designation, range estimation, fire orders, and setting of sights. Every effort is made to carry out the fundamentals of concealment, camouflage, and scouting and patrolling in the conduct of these exercises. Soldiers acting as personnel targets should be rotated with gunners taking the course, and type targets should be shifted frequently to avoid monotony.

(1) *No. 1.*—A stretch of terrain not to exceed 400 yards in length and containing as many natural features as possible, such as trees, shrubs, tall grass, ditches, and walls, is selected for the course. Along a designated path are placed actual personnel targets at various ranges from the path and in normal concealment. Typical targets should include prone, kneeling, and standing soldiers, individual and groups of moving men, machine guns with normal crews, and mounted scouts. The gunner is required to proceed down the designated path and locate targets, estimate ranges, set sights, take position, and simulate fire on each target that he locates. He is accompanied by an instructor who checks all phases of the gunner's action on each type target.

(2) *No. 2.*—In a suitably selected location, a road block of any type should be established and held by a detachment of machine guns and riflemen. Either the submachine gunner dismounted, the motorcyclist with submachine gun, an armored vehicle with submachine gun as an alternate weapon, or any combination of these, should operate against the road block. A commissioned officer should accompany the individual or the vehicle to check and instruct in procedure and criticize the commands of the individual or car commander during the problem.

(3) *No. 3.*—A small area in which buildings predominate and which can be presumed to be a village or city street should be selected for this exercise. The area should be such as to allow personnel to occupy buildings, on roofs and at windows, and to erect barricades. Personnel armed with the submachine gun and mounted on motorcycles and in armored

cars should be required to operate against personnel in buildings and to reduce the barricade.   This type of problem is especially beneficial in the training of mounted and dismounted action, collective firing with other weapons, and proper leadership.   All actions by individuals, squad or platoon leaders, and individual units should be carefully checked by a commissioned officer, and the exercise should be reviewed and criticized immediately upon completion.

CHAPTER 7

ADVICE TO INSTRUCTORS

■ 95. Purpose.—The provisions of this chapter are to be accepted as a guide. They have not the force of regulations. They are particularly applicable to emergency conditions when large bodies of troops are being trained under officers and noncommissioned officers who are not thoroughly familiar with approved training methods.

■ 96. Mechanical Training.—a. As a general rule instruction is so conducted as to insure the uniform progress of the platoon and company.

b. The instructor briefly explains the subject to be taken up and demonstrates it himself or with a trained assistant.

c. The instructor then causes one man in each squad or subgroup to perform the step while he again explains it.

d. The instructor next causes all members of the squads or subgroups to perform the step, checked by their noncommissioned officers. This is continued until all men are proficient in the particular operation, or until those whose progress is slow have been placed under special instructors.

e. Subsequent steps are taken up in like manner during the instruction period.

■ 97. Marksmanship.—a. General.—Training is preferably organized and conducted as outlined in chapters 3, 4, and 5. Officers should generally be considered as the instructors of their units. As only one step is taken up at a time, and as each step begins with a lecture and a demonstration showing exactly what to do, the trainees, although not previously instructed, can carry on the work under the supervision of the instructor.

b. Instructors.—It is advantageous to have all officers and as many noncommissioned officers as possible trained in advance in the prescribed methods of instruction. When units are undergoing marksmanship training for the first time, this is not always practicable nor is it absolutely necessary. A good instructor can give a clear idea of how to carry on the work in his lecture and demonstration preceding each step. In the supervision of the work following the demonstration, he can correct any mistaken ideas or misinterpretations.

c. Equipment.—The instructor should personally inspect the equipment for the preparatory exercises before the training begins. A set of model equipment should be prepared in advance by the instructor for the information and guidance of the organization about to take up the preparatory work.

d. Inspection of submachine gun.—No man is required to fire with an unserviceable or inaccurate weapon. All submachine guns are carefully inspected far enough in advance of the period of training to permit organization commanders to replace all inaccurate or defective weapons.

e. Ammunition.—The best ammunition available is reserved for record firing, and the men should have a chance to learn their sight settings with that ammunition before record practice begins. Ammunition of different makes and of different lots should not be used indiscriminately.

f. Vehicles and drivers.—The best vehicles and drivers of each organization are made available when vehicular firing begins. Vehicles are thoroughly inspected mechanically and suitable for the type firing desired.

g. Ranges.—All ranges to be used, including those for moving targets and air targets, are carefully inspected far enough in advance of the period of use to permit changes or repairs when necessary. Targets and other equipment will be in the best state of repair possible when range practice begins. Arrangements for firing at towed air targets are made with the proper authorities in sufficient time to insure the necessary targets and towing airplanes at the proper time.

71

■ 98. ORGANIZATION OF THE WORK IN MARKSMANSHIP.—*a. In preparatory training.*—The field upon which the preparatory work is to be given is selected in advance and a section of it assigned to each organization. The equipment and apparatus for the work will be on the ground and in place before the lecture is given, so that each organization can move to its place and begin work immediately and without confusion.

b. In range practice.—(1) The range work is so organized that there is a minimum of lost time on the part of each man. Long periods of inactivity while awaiting a turn on the firing line will be avoided. For this purpose no more men should be on the range at one time than the number of targets or ranges available can accommodate efficiently.

(2) In moving ground and air target firing and vehicular firing, it is advisable to have on the line the order that is next to fire and to have them practice with dummy ammunition or simulated fire. When the size of the firing point makes this action impracticable, each order should be given a score of simulated fire before firing with ball cartridges.

■ 99. LECTURES AND DEMONSTRATIONS IN MARKSMANSHIP.—*a.* The lectures at the beginning of each step are an important part of the instructional methods. The lectures may be given to the assembled command or group.

b. The lecturer will know in advance what he is going to say on the subject. Under no circumstances will he read to a class an outline or lecture prepared by himself. If the instructor cannot talk interestingly and instructively on each subject without the use of copious notes, he should not be giving the lectures at all.

c. The one important thing is to show the men undergoing instruction, by explanation and demonstration, just how to go through the exercises and to impress upon them why they are given.

■ 100. TECHNIQUE OF FIRE; RANGE ESTIMATION.—*a. General.*—The instructor will secure the necessary equipment, lay out and supervise the construction of the courses to be run, and detail and train the necessary assistants prior to the first period of instruction. Instructors use their initiative in arranging additional exercises to those given in paragraph 94. It will be explained to pupils how the exercises used illustrate the principles in the technique of fire. Good work in the conduct of the exercises as well as errors will be called to the attention of all pupils.

b. Range estimation.—The instructor makes the necessary arrangements to set up the training courses for range estimation given in paragraph 93 prior to the first period of this training. All equipment such as paper and pencils will be available to each group of pupils at the beginning of each instruction period.

■ 101. ALLOTMENT OF TIME.—A suggested allotment of time for conducting range practice is tabulated below. The time indicated is for a troop or similar organization rather than for individual firing.

Subject: *Time (hours)*
    Instruction, dismounted practice course_____ 9
    Instruction, vehicular firing_____ 12
    Instruction, moving ground and air targets__ 12
    Record, dismounted practice course_____ 3

■ **102. RECORD OF INSTRUCTION AND QUALIFICATION.**—A convenient form for keeping a record of the status of each individual's training is shown below:

RECORD OF SUBMACHINE GUN INSTRUCTION

_____   _____   _____   _____
(Last name)   (First name)   (Middle initial)   (Army serial No.)

Grade _____   Organization _____

Preparatory Training and Gunner's Test

	Date	Result	Initials
General description of weapon			
Removal of groups			
Replacement of groups			
Disassembling of groups			
Assembling of groups			
Care, cleaning, and oiling			
Functioning, including use of safety and fire control			
Stoppages and immediate action			
Spare parts, name and explain use			
Positions, standing, kneeling, prone			
Adjustment and use of sights			
Range estimation, effect of wind and light			
Safety precautions, individual			
Loading and unloading, clip and drum			
Examination			

METHOD OF MARKING (under "Result")

Excellent: √ √   Unsatisfactory: X
Satisfactory: √   Inferior: X X

[Front]

RECORD OF SUBMACHINE GUN INSTRUCTION

------------------- ------------------- ------------------- -------------------
(Last name)   (First name)   (Middle initial)   (Army serial No.)

Grade _____ Organization _____

Marksmanship Training and Record Firing

Marksmanship exercises (par. 48)				Date	Initials
First					
Second					
Third					

Preliminary firing	Targets hit	Number of hits	Score		
Dismounted practice					
Dismounted practice					
Dismounted practice					
Vehicular firing					
Vehicular firing					
Moving ground targets					
Air target					

Record firing					Signature
Dismounted practice					

Qualification (dismounted practice course):
    Expert.
    1st class gunner.
    2d class gunner.

(Signature of organization commander.)

(Grade and organization.)

[Back]

INDEX

77

INDEX

INDEX

○

TECHNICAL MANUAL }
No. 9–1215   }

WAR DEPARTMENT,
WASHINGTON, March 1, 1942.

ORDNANCE MAINTENANCE, THOMPSON SUBMACHINE GUN, CAL. .45, M1928A1

Prepared under direction of the
Chief of Ordnance

SECTION I

INTRODUCTION

1. Scope.—This manual is published for the information and guidance of ordnance maintenance personnel. It contains detailed instructions for inspection, disassembly, assembly, maintenance, and repair of the Thompson submachine gun, cal. .45, M1928A1, supplementary to those in the Field and Technical Manuals prepared for the using arm. Additional descriptive matter and illustrations are included to aid in providing a complete working knowledge of the matériel.

SECTION II

GENERAL DESCRIPTION

2. Description.—The Thompson submachine gun, cal. .45, M1928A1 (figs. 1 and 2), is an aircooled, blowback-operated, magazine-fed weapon. It is designed to be fired from the shoulder of the gunner similarly to a rifle, and is used as an auxiliary weapon by the United States Army, Navy, and Marine Corps. The fire control lever of the gun can be set for either full automatic or semiautomatic

RA PD 4106

FIGURE 1.—Thompson submachine gun, cal. .45, M1928A1— right side view with 20-round magazine.

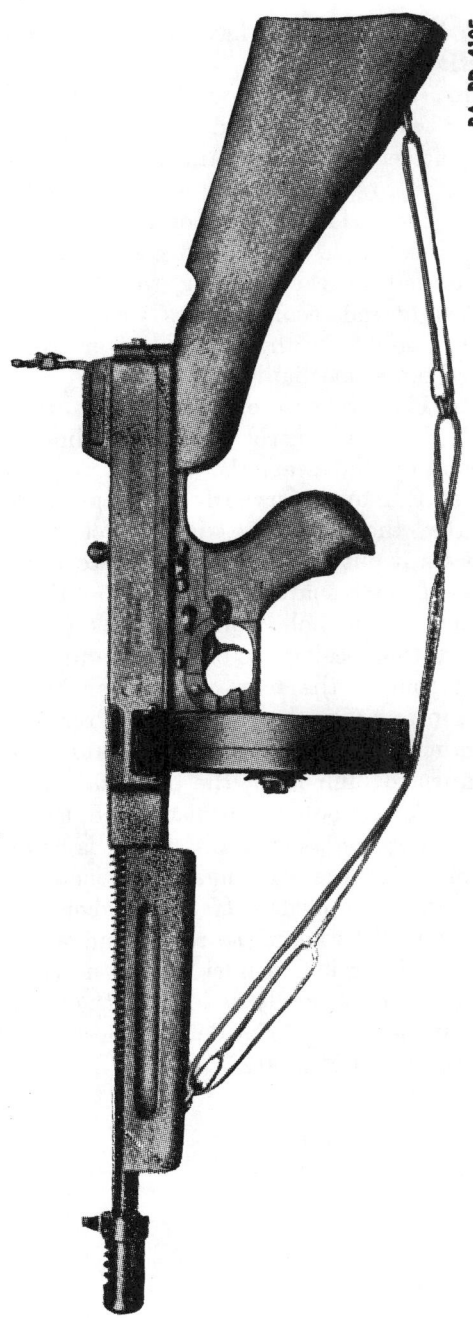

RA PD 4105

Figure 2.—Thompson submachine gun, cal. .45, M1928A1— left side view with 50-round magazine.

fire. Two types of magazines, having capacities of 20 and 50 cartridges respectively, may be used.

3. Mechanism (fig. 3).—The gun is composed of two distinct groups, the frame with its attached and contained parts, and the receiver with its attached and contained parts. The frame group consists of the frame, to which are attached the butt stock assembly and rear wooden grip. The frame contains the trigger and sear groups, the magazine catch, and the fire control mechanism. The receiver group consists of the receiver, to which are attached the barrel with front sight and recoil ("Cutts") compensator, the wooden foregrip and rear sight. Within the receiver are the bolt, lock, actuator, recoil and buffer mechanisms.

4. Operation.—The cycle of operation is as follows: With the safety at "Fire," the fire control lever at "Single" and the bolt retracted and held by the sear, the trigger is pulled. The bolt, released by the sear, moves forward under pressure of the recoil spring. The end of the bolt comes in contact with the base of a cartridge and forces it out of the magazine into the chamber of the barrel, where the extractor snaps over the rim of the cartridge. The forward movement of the bolt cams the lock downward into the locking grooves of the receiver so the bolt and receiver are completely locked together in the forward position before the hammer forces the firing pin to strike the cartridge. Pressure of the exploding cartridge against the end of the bolt, transmitted to the lock, forces the lock upward, unlocking the bolt and drives it backward with the actuator. As the bolt moves backward, the empty cartridge case is extracted and ejected, the recoil spring is compressed against the buffer pilot collar and the sear engages in one of the two notches of the bolt, completing the cycle. If the fire control lever is set at "Full Auto," the sear will remain depressed and will not engage the bolt on the backward stroke. Under this condition, the gun will continue to function automatically as long as the trigger is retracted or until the magazine is empty. For detailed description of operation and functioning, refer to FM 23–40.

RA PD 9538

FIGURE 3.—Thompson submachine gun, cal. .45, M1928A1—groups.

SECTION III

INSPECTION

5. General.—*a.* Inspection is for the purpose of determining the condition of the matériel, whether repairs or adjustments are required, and the remedies necessary to insure that the matériel is in serviceable condition.

b. Before inspection is begun, the equipment should be thoroughly cleaned to remove any fouling, dirt or other foreign matter, which might interfere with its proper functioning. For instructions in care and cleaning, and materials used, refer to FM 23–40, section on "Care and Cleaning"; section IV of this Technical Manual, TM 9–850, and SNL K–1.

6. Inspection report.—The procedure to be followed relating to inspection and maintenance is contained in TM 9–1100, "Ordnance Maintenance Procedure—Matériel Inspection and Repair."

7. Tools for inspection.—Tools used for inspection of the gun are those furnished for disassembly, assembly and repair. They are included in the accessories referred to in FM 23–40, and listed in SNL A–35.

8. Gun as a unit.—*a.* Check gun for general appearance, metal parts for scratches, rust, and wear, and wooden parts for cracks and nicks. Check firmness of magazine in grooves, and action of magazine catch. Check rigidity of rear sight base on receiver, front sight on compensator, compensator on barrel, butt stock and grips on frame and receiver, and sling swivels on stock and foregrip. Inspect heads of screws for burs. Remove magazine and check smoothness of bolt and trigger action while retarding actuator movement by hand so the bolt will not fly forward on an empty chamber.

Caution: Unless magazine is removed, the bolt, if released, will fire a cartridge from a loaded magazine, as this gun fires on forward stroke of the bolt.

b. If possible and practicable, fire several rounds from the gun. Observe action of the weapon and analyze the cause of any malfunction.

RA PD 9540

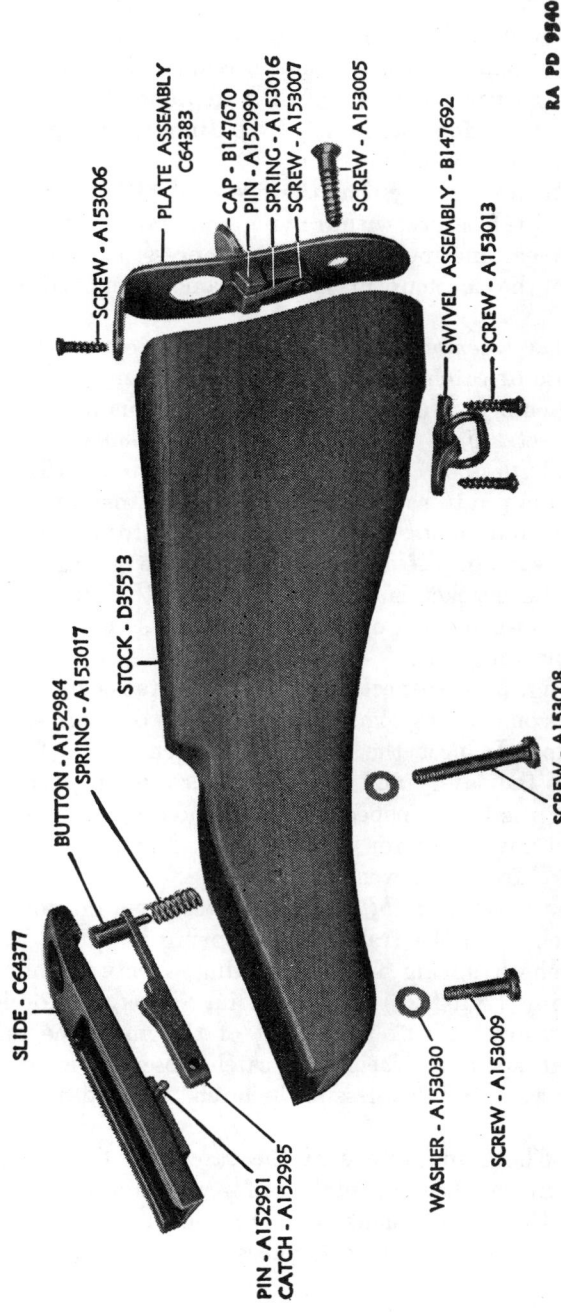

SCREW - A153006

PLATE ASSEMBLY
C64383

CAP - B147670
PIN - A152990
SPRING - A153016
SCREW - A153007

SCREW - A153005

SWIVEL ASSEMBLY - B147692

SCREW - A153013

STOCK - D35513

BUTTON - A152984
SPRING - A153017

SCREW - A153008

SLIDE - C64377

WASHER - A153030

SCREW - A153009

PIN - A152991
CATCH - A152985

FIGURE 4.—Thompson submachine gun, cal. .45, M1928A1—butt stock disassembled.

9. Frame group.—*a.* To inspect the frame group, first remove the butt stock assembly. Pull the actuator to rear until the bolt is caught and held by the sear. With the bolt in rearward position, set the fire control lever (rocker pivot) at "Full Auto" and the safety at "Fire" and allow bolt to go forward slowly by pulling the trigger and retarding the actuator by hand.

Caution: It is necessary that the fire control lever and the safety be set as described before withdrawing the frame from the receiver. Otherwise the sear and rocker will not be depressed and serious damage can result to the mechanism if the frame is moved under these conditions.

b. Butt stock assembly (fig. 4).—Check action of the butt stock catch, and nose of catch for wear and burs. Remove the screws and lift out the assembly. Inspect butt stock catch spring for functioning, fracture and set. Free length of spring (A153017) is .75+.02 in. Drive out the butt stock catchpin and remove catch. (To remove the catch button, file pin to round and drive out.) Inspect the butt stock slide for burs and dents. Remove the butt stock plate to inspect action of trap spring. If necessary, remove the spring and drive out trap pin. (The bracket is riveted to the butt plate.) Remove the sling swivel screws and lift out the swivel plate. If necessary, spring the swivel from the plate.

c. Frame (fig. 5).—Inspect the frame for cracks in the metal and dents and burs on corners, grooves and surfaces of stock slideways and magazine grip. Inspect the butt stock catch notch for wear and burs. Inspect the safety and the pivot holes in the frame for wear. If the rear grip is loose, inspect screw threads in frame and on screw for wear, and the screw for straightness. Inspect the frame latch notch in rear of frame for wear and burs.

d. Magazine catch assembly.—Unless necessary, do not remove the magazine catch from the frame as the spring is apt to be damaged. If removed, check spring for functioning, fracture and set. Free length of spring (A153021) is .85+.02 in. Look for foreign matter in spring aperture. Check movement of the magazine catch in the frame without spring. Check the catch nose for wear and burs. Check pin for wear and firmness in the latch. (Head of pin is riveted into latch.)

e. Safety.—Check movement of the safety in the frame without pivot plate. Inspect bearing surfaces for wear and burs.

f. Rocker.—Check movement of the rocker on rocker pivot. Inspect for wear and burs on contacting surfaces.

RA PD 9539

GRIP - C64372

SCREW - A153011

CATCH ASSEMBLY
C64380

FRAME - D35511

SAFETY
B147683

PIVOT
B147679

SPRING
A153021

SEAR - B147684

SPRING - A153025

SPRING
A153026

LEVER
B147675

PLATE ASSEMBLY - B147690

ROCKER - B147682

TRIP - B147686

DISCONNECTOR - B147671

SPRING - A153018

TRIGGER - C64378

SPRING - A153027

FIGURE 5.—Thompson submachine gun, cal. .45, M1928A1—frame group disassembled.

g. Rocker pivot.—Check movement of the rocker pivot in the frame without pivot plate. Inspect bearing surfaces for wear and burs.

h. Pivot plate assembly.—Check the pivot plate short and long springs for functioning, fracture, and set. Inspect pivot pins for wear and firmness in the plate. (Pin heads are riveted into plate.)

i. Sear group (fig. 5).—Check sear for retention of the bolt. Inspect the sear for wear and burs on nose and contracting surfaces, and for foreign matter in spring aperture. Check movement of the sear with sear lever on the pivot. Check sear spring for functioning, fracture, and set. Free length of spring (A153025) is .765+.02 in. Inspect sear lever for wear and burs on contacting surfaces and for foreign matter in spring aperture. Check sear lever spring for functioning, fracture, and set. Free length of spring (A153026) is .43±.01 in.

j. Trigger group (fig. 5).—Inspect the trigger for wear in pivot hole and disconnector pivot hole, and for deformation of tip. Check action of trigger with disconnector and trip on trigger pivot. Check trigger spring for functioning, fracture, and set. Free length of spring (A153027) is .65±.01 in. Inspect disconnector for wear on pin and wear and burs on contacting surface. Check disconnector spring for functioning, fracture, and set. Free length of spring (A153018) is .50±.01 in. Inspect trip for wear and burs on contacting surfaces.

k. Assemble the sear and trigger mechanisms in the frame and try action of the trigger with the rocker in each position.

10. Receiver and barrel groups (fig. 6).—*a.* To inspect the receiver group, remove the buffer pilot and pad together with the recoil spring from the receiver. With the receiver inverted, move the actuator back and forth to inspect movement of the lock in the grooves of receiver and bolt. Take out the bolt group, actuator and lock, and remove oiler assembly.

b. Receiver group.—Inspect the receiver for wear and burs on frame slideways and lock camming surfaces, and the bullet ramp for wear and fouling. Inspect corners and edges for dents and burs and the magazine retaining grooves and actuator groove for wear. Inspect the buffer pilot aperture for wear. Inspect the frame latch for wear and aperture for wear and foreign matter. Check latch spring for functioning, fracture, and set. Free length of spring (A153020) is .45±.01 in. Check the ejector for firmness in receiver, and inspect the point for wear and alinement. Never try to remove the ejector from receiver with the bolt in the forward position.

c. Barrel group.—Inspect the barrel as a unit from the standpoint of serviceability.

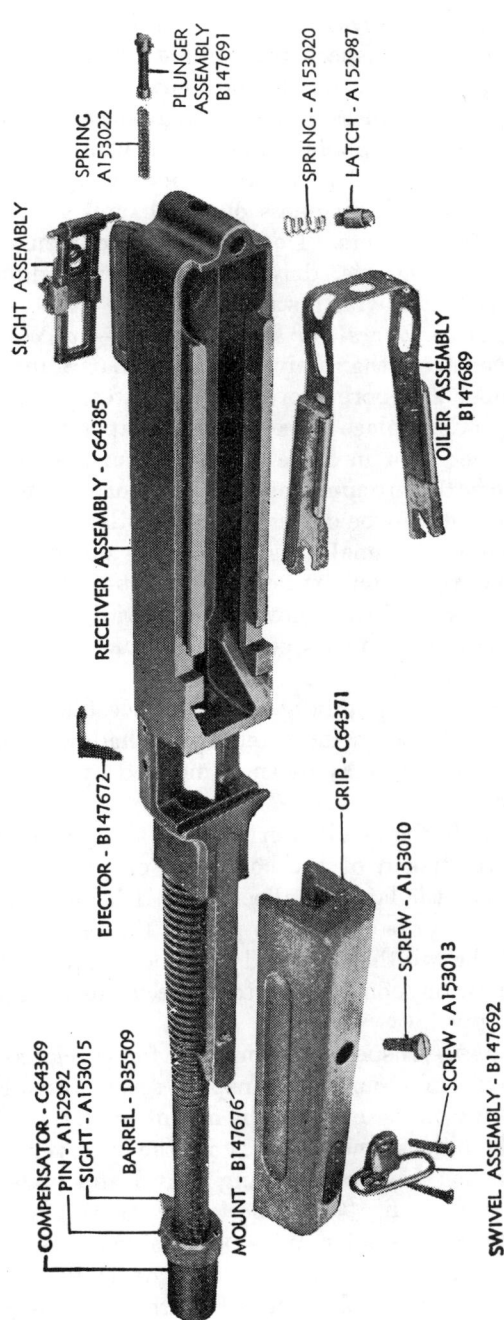

RA PD 9541

SIGHT ASSEMBLY

PLUNGER ASSEMBLY B147691

SPRING A153022

SPRING - A153020

LATCH - A152987

RECEIVER ASSEMBLY - C64385

OILER ASSEMBLY B147689

EJECTOR - B147672

GRIP - C64371

SCREW - A153010

SCREW - A153013

COMPENSATOR - C64369

PIN - A152992

SIGHT - A153015

BARREL - D35509

MOUNT - B147676

SWIVEL ASSEMBLY - B147692

FIGURE 6.—Thompson submachine gun, cal. .45, M1928A1— receiver and barrel group disassembled.

11

(1) *Inspection of the barrel group as a unit.*—Check firmness of the barrel in receiver. Inspect the annular radiator grooves for presence of foreign matter, dents and burs. Do not remove the barrel from the receiver unless necessary to replace. For removal of the barrel, refer to paragraph 14 *c* (2) (*b*). Inspect for loose front sight, alinement of sight blade, burs and shine on tip. Inspect recoil ("Cutts") compensator for firmness on the barrel and for any foreign matter in gas escape slots. Do not remove compensator or sight unless necessary. To remove, drive out pin, drive sight rearward and unscrew compensator with a strap wrench.

(2) *Inspection of the barrel for serviceability.*—(*a*) With the firing mechanism removed from the receiver, hold the barrel up to the light, and inspect chamber and bore thoroughly for wear, pits and bulges. To facilitate inspection, place piece of white paper or rag in the receiver so as to reflect light into the bore, then turn the barrel slowly so the light follows the circumference of the bore. Untrueness and bulges in the bore can thus be detected more easily.

(*b*) A barrel containing small pits, but having sharp and uniformly distinct lands, and free from bulges, will be sufficiently accurate to be serviceable. This condition, however, naturally implies that the barrel has been neglected and its period of serviceability will, therefore, be materially lessened.

(*c*) A barrel containing a bulge is unserviceable and should be scrapped. This condition is indicated by a shadowy depression or dark ring in the bore and may often be noticed by a raised ring on the barrel surface.

(*d*) A barrel with the lands worn away for a considerable distance from the breech end of the bore, and/or pitted to the extent that the sharpness of the lands is affected, or if it has a pit or pits in the lands or grooves large enough to permit the passage of gas past the bullet (a pit the width of a land or groove and ⅜ to ½ in. in length or longer) is, or soon will be, too inaccurate for serviceability and should be scrapped.

d. Foregrip group.—Inspect the foregrip for cracks and rigidity. Remove foregrip from mount and inspect screw threads on screw and in the mount for burs. Remove the mount by driving it forward, then inspect slide blades and grooves in the receiver. Check the mount for alinement. Inspect the sling swivel and plate.

e. Rear sight group (fig. 7).—Check for missing or loose base rivets in the receiver. Do not remove base unless necessary. Inspect base for breaks and dents. Check action of sight leaf assembly with plunger and spring. Inspect all parts for rust, alinement, deformation and presence of foreign matter. Check operation of windage screw and sight slide and sight slide catch. Do not remove unless necessary. (To remove, drive out sight base pin and remove sight leaf assembly.) Remove plunger and plunger spring. Inspect point of plunger for wear, plunger pin for deformation and spring for functioning, fracture, and set. Free length of spring (A153022) is $1.20 \pm .02$ in. Check for missing or loose sight slide stop pin. Do not disassemble unless necessary. (To disassemble, drive out pin and remove slide. Drive out windage screw collar pin and remove collar

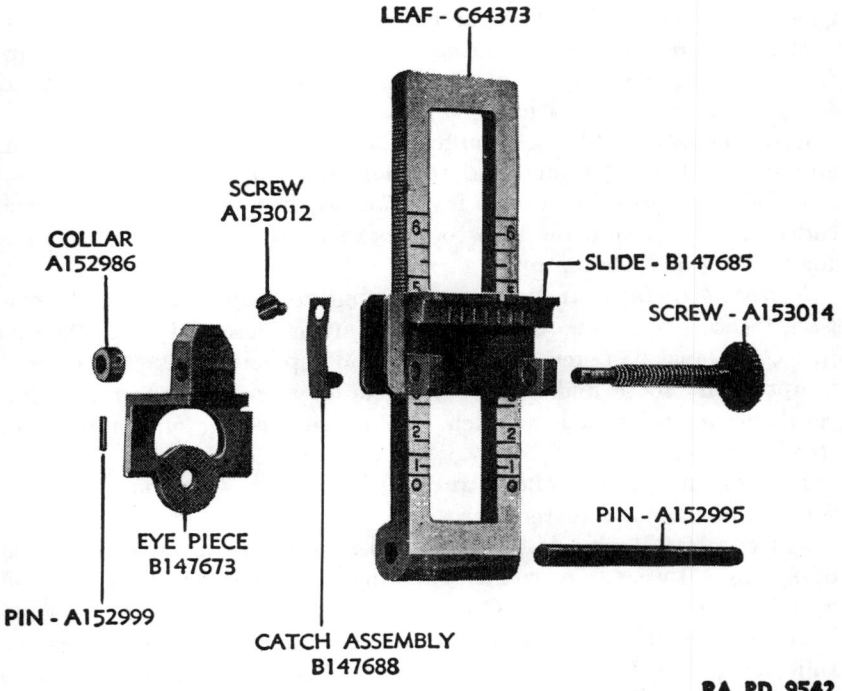

RA PD 9542

FIGURE 7.—Thompson submachine gun, cal. .45, M1928A1—rear sight group disassembled.

and windage screw and the eye piece.) Inspect the eye piece slides and the slide grooves for burs and wear.

f. Bolt group (fig. 8).—Bolt surfaces should be polished and free from rust, foreign matter, or roughness. Inspect sear notches, edges, corners and grooves for burs, wear and dents. Examine fired cartridge cases for indications of set back primer due to worn face of bolt or enlarged firing pin hole. Inspect the bolt face for deformation and firing pin hole for enlargement. Inspect head of T-slot for burs and wear caused by the rocker. Remove the extractor, taking care not to spring more than necessary to clear the lug. Inspect the extractor for set and deformation, and claw for wear. Remove the hammer pin, hammer, firing pin and firing pin spring. Check hammer pin for looseness and wear. Check hammer for wear and burs in pin hole and on contacting surfaces. Inspect firing pin head and nose for wear and burs. Check firing pin spring for functioning, fracture, and set. Free length of spring (A153019) is 2.50+.05 in.

g. Lock.—Inspect lock for wear and burs on sliding surfaces. Check movement of lock in actuator and receiver locking grooves.

h. Actuator.—Inspect actuator for wear and burs on sliding surfaces, recoil spring aperture for foreign matter and actuator head for fracture or deformation.

i. Buffer group.—Inspect buffer pilot for alinement, deformation, and wear. Inspect buffer pad for deformation and wear.

j. Oiler group.—Check oiler for fit and spring retention in receiver. Sides of oiler should lie flush to sides of receiver. Check oiler pads for fraying and absorption.

k. Recoil spring.—Inspect recoil spring for functioning, deformation, fracture, and set. Free length of spring (A153024) is 10.00+.25 in. Care must be taken in removing and replacing this spring, as it is apt to fly loose and become twisted between actuator and pilot, resulting in deformation which may cause binding on compression stroke of bolt.

11. Box magazine (20 rounds) (fig. 9).—*a.* Check box magazine for fit and retention in receiver.

b. Depress follower and note smoothness of operation and tension of spring. Insert two or three dummy cartridges in magazine and attach magazine to gun. Operate the piece by hand and observe loading, extraction, and ejection. Note also whether the magazine follower (when the magazine is empty) lifts the trip sufficiently to force the disconnector from under the sear lever and allow the sear to catch the bolt and hold it in the open position.

c. Inspect magazine tube for dents, cracks, deformed lips, and foreign matter. Check follower pin for deformation, wear and burs, and magazine spring for functioning, fracture, and set. Free length of spring is 8.00+.20 in.

12. Drum magazine (50 rounds) (fig. 10).—*a.* Check drum magazine for fit and retention in receiver.

b. Remove winding key and note its condition. Remove cover and check cover guide, slide and rivets. Rotate the rotor, noting the

RA PD 9543

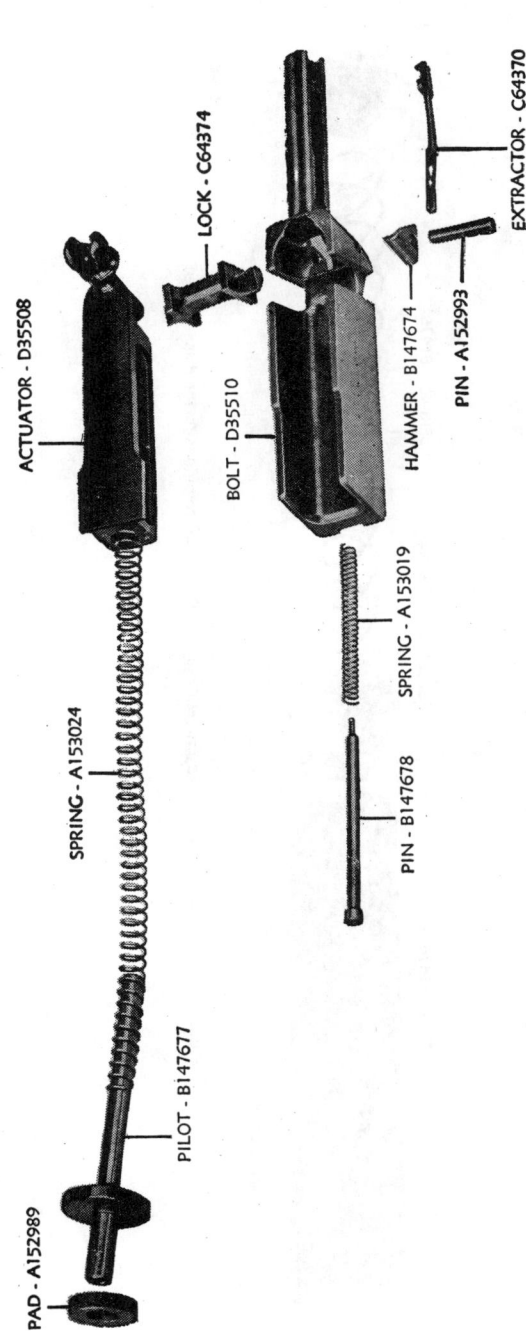

LOCK - C64374

EXTRACTOR - C64370

ACTUATOR - D35508

HAMMER - B147674

PIN - A152993

BOLT - D35510

SPRING - A153019

SPRING - A153024

PIN - B147678

PILOT - B147677

PAD - A152989

FIGURE 8.—Thompson submachine gun, cal. .45, M1928A1—bolt group disassembled.

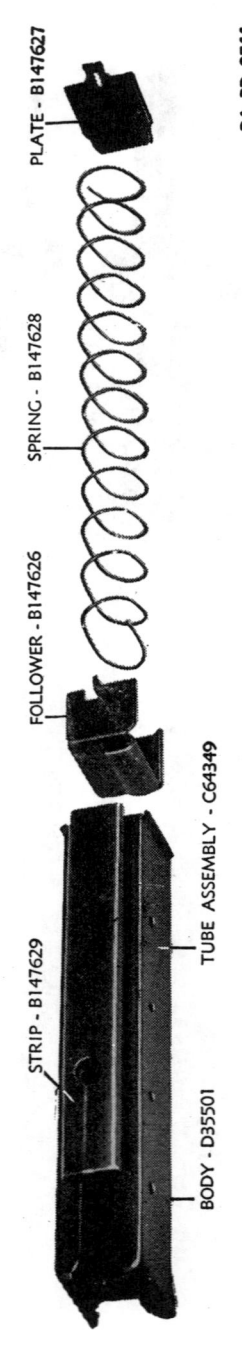

PLATE - B147627

SPRING - B147628

FOLLOWER - B147626

TUBE ASSEMBLY - C64349

STRIP - B147629

BODY - D35501

RA PD 9544

FIGURE 9.—Thompson submachine gun, cal. .45, M1928A1—20-round magazine disassembled.

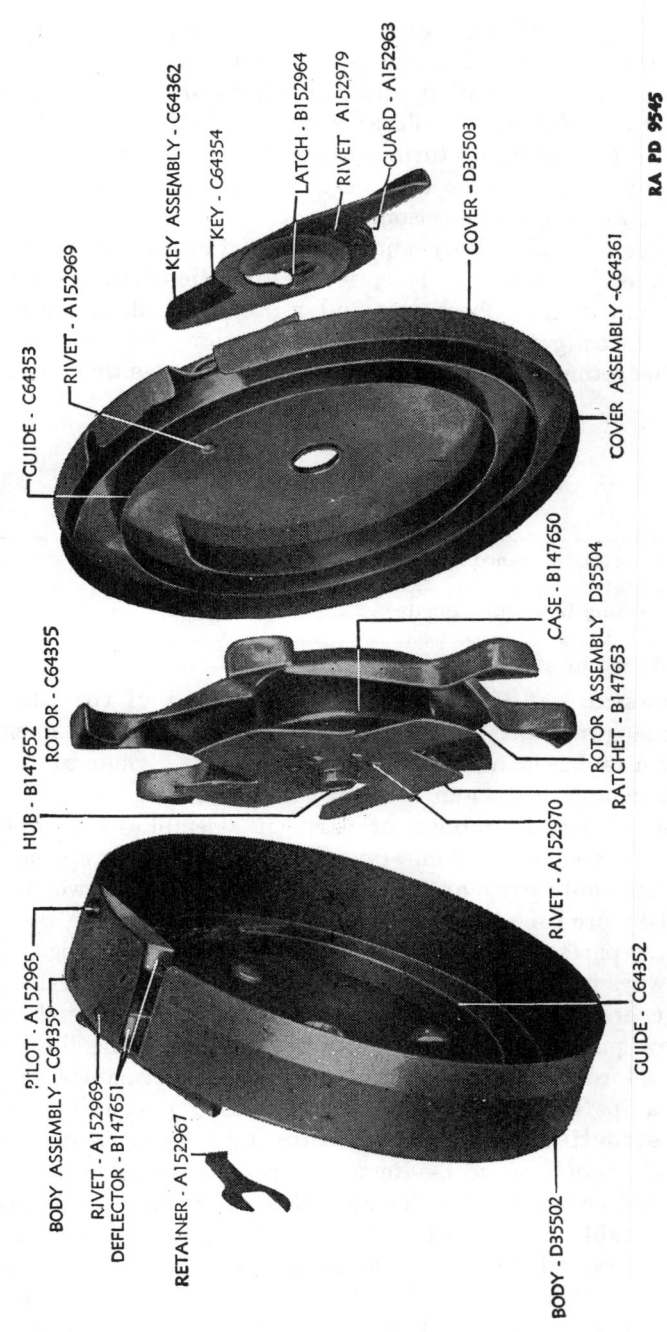

KEY ASSEMBLY - C64362
KEY - C64354
LATCH - B152964
RIVET - A152979
GUARD - A152963
COVER - D35503
COVER ASSEMBLY - C64361
RIVET - A152969
GUIDE - C64353
CASE - B147650
ROTOR ASSEMBLY D35504
RATCHET - B147653
HUB - B147652
ROTOR - C64355
RIVET - A152970
PILOT - A152965
BODY ASSEMBLY - C64359
RIVET - A152969
DEFLECTOR - B147651
RETAINER - A152967
GUIDE - C64352
BODY - D35502

RA PD 9545

FIGURE 10.—Thompson submachine gun, cal. .45, M1928A1—50-round magazine disassembled.

action of the rotor spring. Further test the magazine by forcing the finger of the rotor standing nearest the feedway in the opposite direction to its normal rotation, and insert five dummy cartridges (one resting in the feedway). Replace cover and winding key and increase tension of rotor spring by turning the winding key one click. Insert the magazine and operate the piece rapidly by hand and observe loading, extraction, and ejection.

c. Inspect magazine body and cover for deformation and dents. Check guides in cover and body for deformation, and deflector for looseness and wear. Check key and retainer for deformation, and magazine catch grip in body for deformation and burs. Do not disassemble rotor and spring case unless necessary, as they are riveted together.

Section IV

MAINTENANCE AND REPAIR

13. General.—*a.* The maintenance and repair of the Thompson submachine gun, cal. .45, M1928A1, as covered in this manual is primarily a replacement of worn or broken parts. General disassembly and assembly of the gun is covered in FM 23–40.

b. Where parts, assemblies, or parts of assemblies are broken or worn so as to render them unserviceable, they must be replaced from stock. Often only parts of assemblies will be broken or worn; where it will take more time to remove the serviceable parts from the assembly than the parts are worth, the assembly should be scrapped. Parts do not always interchange and should be assembled by selection.

c. In general, maintenance operations are of a first aid nature, performed by qualified ordnance personnel with only the limited tool facilities afforded by repair trucks, or by semipermanent shops at posts or camps, or by an inspector while making a regular inspection.

14. Instructions for maintenance and repair.—*a. Burs on screws and smooth surfaces.*—Remove burs from screw heads, threads and like surfaces with a fine file, and chase out damaged threads with a die if available. Remove the burs from smooth contacting surfaces with a fine grained sharpening stone or emery cloth, and finish with crocus cloth. Polish rounded contacting surfaces with crocus cloth. Care should be observed to file and stone evenly and lightly, removing no more metal than is necessary. For materials employed in removing rust, cleaning and preserving, and the limits of their proper use, refer to TM 9–850.

b. Frame group.—(1) *Butt stock, assembly.*—If the butt stock is not held rigidly to the frame by the slide or catch, the faulty parts should be replaced.

(2) *Frame.*—When the frame is damaged to the extent that improper functioning of the gun results, it should be replaced.

(3) *Magazine catch, assembly.*—If the magazine catch does not hold in magazine firmly, it should be replaced. Check to see if the fault lies in the magazine.

(4) *Rocker and rocker pivot.*—The rocker or rocker pivot should be replaced if worn to the extent that automatic firing occurs with the rocker pivot set at "Single."

(5) *Sear, trigger and pivot plate groups.*—(a) When the bearing surfaces on the sear, trigger and the pivot plate pins become worn to the extent that malfunctioning of the gun results, the worn part or parts should be replaced.

(b) If either spring finger on the pivot plate becomes set or broken, replace the pivot plate.

c. Receiver and barrel groups.—(1) *Receiver group.*—(a) A receiver damaged to the extent that malfunctioning of the gun results should be replaced.

(b) A worn ejector should be replaced.

(c) If the frame latch or aperture becomes worn so that the frame is not securely locked to the receiver, the latch or receiver should be replaced.

(2) *Barrel group.*—(a) If it is determined that the barrel is unserviceable by inspection as prescribed in paragraph 10 c (2), the barrel should be replaced.

(b) To remove barrel, disassemble the gun, wedge a block of hard wood in receiver to prevent springing of the side, clamp receiver in a vise with leather jaws and unscrew barrel from receiver, using a strap wrench. If barrel is to be scrapped, a pipe wrench may be used.

(c) When it is determined to replace the barrel, the recoil compensator and front sight, if in good condition, should be removed from the defective barrel for assembly to the new barrel.

(3) *Rear sight group.*—If the rear sight has been broken or bent out of line, the damaged parts or the entire leaf assembly should be replaced.

(4) *Bolt group.*—(a) If the face of the bolt shows signs of wear, or firing pin hole has become enlarged, the bolt should be replaced.

(*b*) The extractor should be replaced if it becomes deformed and does not extract the cartridge properly.

(*c*) If the nose of the firing pin becomes worn or deformed, the firing pin should be replaced.

(5) *Lock.*—It is of extreme importance that the lock be in good condition with all sliding surfaces smooth and polished, otherwise repair or replacement is necessary.

(6) *Oiler group.*—If the oiler is deformed so as to interfere with action of recoiling parts, it should be replaced. If the oil pads are dirty or do not absorb oil properly, replace the oiler.

(7) *Buffer group and recoil spring.*—(*a*) If the buffer pilot or the pad should be deformed to the extent of hindering proper functioning of the gun, they should be replaced.

(*b*) If the recoil spring is kinked or set, it should be replaced.

d. Magazines, box and drum type.—If the springs are weak, they must be replaced. In the drum magazine, the entire rotor should be replaced. If the magazines are deformed so they will not lock in the gun properly or prevent proper action of the spring, the faulty part should be replaced. If the lips are bent or out of true or deformed so they do not feed cartridges to the gun properly and cannot be repaired, the part should be replaced.

15. Care and cleaning.—*a.* It is of great importance that the matériel be kept absolutely clean and ready for inspection or use at all times. Special attention should be given to dirty magazines. After firing, clean the bore, chamber and all parts, and surfaces of the receiver, bolt ejector, and extractor that have come in contact with powder gases. Remove the frame from the receiver and take out the bolt, and thoroughly clean front end of the bolt and the extractor. With the bolt removed, the bolt well, throat of the receiver, and ejector head are readily accessible.

b. The bore is best cleaned with Cleaner, rifle bore, as prescribed in TM 9–850 in sections entitled "Cleaners and Preservatives," and "Lubricants." When rifle bore cleaner is not available, soap and water should be used as prescribed in FM 23–40. For material used in care and preservation of the gun, refer to TM 9–850 and SNL K–1.

16. Care and cleaning in Arctic climates.—For special care and cleaning of the gun in Arctic climates, refer to TM 9–850, section on "Lubricants."

17. Lubrication.—The gun should be kept thoroughly lubricated at all times. The felt pads in the breech oiler should be kept well saturated with oil. However, the oil contained and distributed by the felt pads is not sufficient in instances of prolonged firing, so all sliding surfaces should be oiled frequently and freely to insure perfect functioning of the gun. For proper instruction in the lubricating of the gun, refer to FM 23–40, and for material used, TM 9–850 and SNL K–1.

18. Matériel affected by gas.—For defense against chemical attack, and for procedure to be followed in the care of matériel affected by gas, refer to FM 21–40 and TM 9–850.

Section V

REFERENCES

19. Standard Nomenclature Lists.

a. Ammunition, revolver and automatic pistol_____ SNL T–2

b. Cleaning and preserving.

     Cleaning, preserving and lubricating material, recoil fluids, special oils, and similar items of issue_____ SNL K–1

     Soldering, brazing, and welding materials, and related items_____ SNL K–2

c. Gun matériel.

     Gun, submarine, cal. .45, Thompson, M1928A1_ SNL A–32

     Tools, special repair, automatic guns, automatic gun matériel, automatic and semiautomatic cannon and mortars_____ SNL A–35

     Truck, small arms, repair, M1_____ SNL G–72

     Current Standard Nomenclature Lists are as tabulated here. An up-to-date list of SNL's is maintained as the "Ordnance Publications for Supply Index"_____ OPSI

20. Explanatory publications.

a. Ammunition, general_____ TM 9–1900

b. Cleaning, preserving, lubricating, and welding materials, and similar items issued by the Ordnance Department_____ TM 9–850

c. Gun matériel.

     Defense against chemical attack_____ FM 21–40

     Ordnance maintenance procedure—Matériel inspection and repair_____ TM 9–1100

     Thompson submachine gun, cal. .45, M1928A1__ FM 23–40

INDEX TO TEXT

23

[A. G. 062.11 (12–23–41).]

By order of the Secretary of War:

G. C. MARSHALL,
Chief of Staff.

Official:
E. S. ADAMS,
Major General,
The Adjutant General.

Distribution: Bn 9 (1) ; IC 9 (3).
(For explanation of symbols see FM 21–6.)

THOMPSON
SUBMACHINE GUN
CAL. .45, M1

WAR DEPARTMENT • *10 OCTOBER 1942*

WAR DEPARTMENT

Washington 25, D. C., 10 October 1942

TM 9-215, Thompson Submachine Gun, Cal. .45, M1, is published for the information and guidance of all concerned.

$\left[\text{A.G. 062.11 (9/9/42)}\right]$

BY ORDER OF THE SECRETARY OF WAR:

G. C. MARSHALL,
Chief of Staff.

OFFICIAL:
J. A. ULIO,
Major General,
The Adjutant General.

DISTRIBUTION: **X.**

(For explanation of symbols, see **FM 21-6.**)

CONTENTS

RESTRICTED

THOMPSON SUBMACHINE GUN, CAL. .45, M1

Section I

INTRODUCTION

1. PURPOSE AND SCOPE.

TM 9-215, dated October 10, 1942, is intended to serve temporarily (pending the publication of a revision now in preparation which will be wider in scope) for the information and guidance of the personnel of the using arms charged with the operation and maintenance of this materiel.

2. CONTENT AND ARRANGEMENT OF THE MANUAL.

Sections I through VIII contain information chiefly for the guidance of operating personnel. Sections IX through XI contain information intended chiefly for the guidance of personnel doing maintenance work.

3. REFERENCES.

Section XII lists all Standard Nomenclature Lists, Technical Manuals, and other publications for the materiel described herein.

INTRODUCTION

RA PD 50287

Figure 1 — Right Side View with 20-Round Magazine Attached

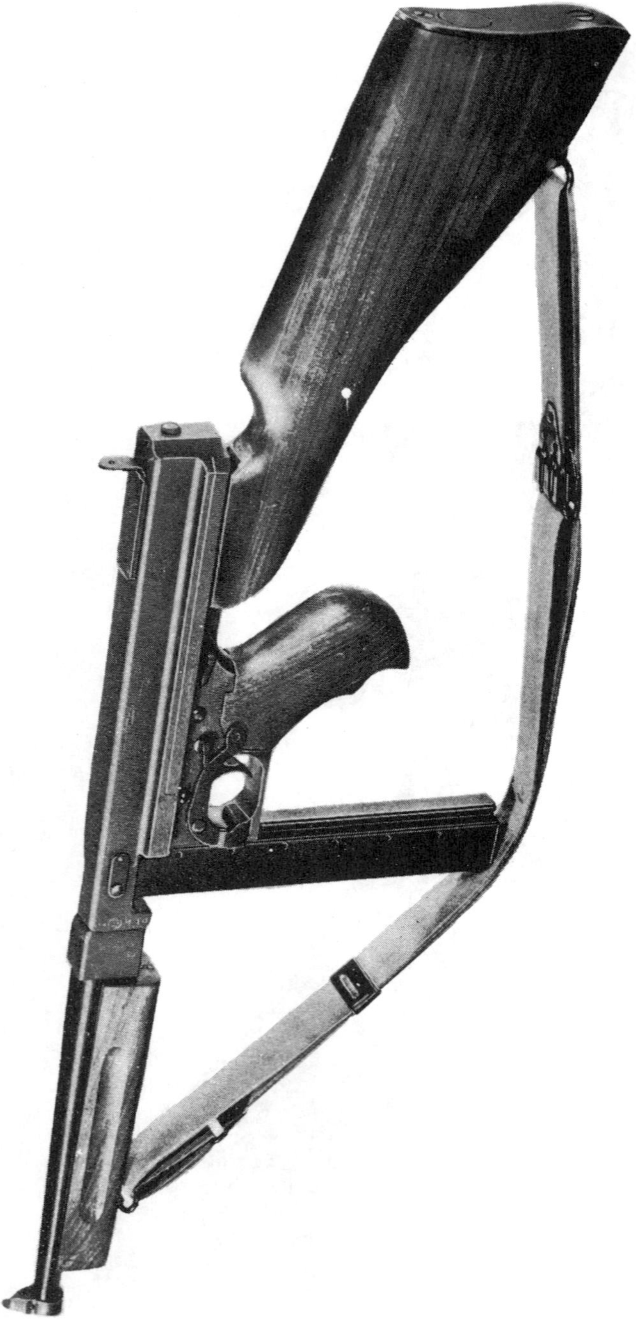

THOMPSON SUBMACHINE GUN, CAL. .45, M1

RA PD 50288

Figure 2 — Left Side View with 30-Round Magazine Attached

4

Section II

DESCRIPTION, DATA, AND CAUTIONS

Paragraph

4. DESCRIPTION.

The Thompson Submachine Gun, Cal. .45, M1 (figs. 1 and 2) is an air-cooled, blowback operated, magazine-fed weapon. It is designed to be fired from the shoulder of the gunner similarly to a rifle. The hand of the gunner is protected on the under side of the barrel by a wooden fore grip; a rear grip is also provided. Sling swivels are secured to the fore grip and stock for attaching the gun sling. By correctly setting the rocker pivot, the weapon may be used for either full automatic fire or semiautomatic fire. Two magazines having capacities of 20 and 30 rounds respectively may be used (figs. 1 and 2).

5. DATA.

Weight of submachine gun (approx.)	10½	pounds
Length of submachine gun (approx.)	32	inches
Weight of 20-round magazine	6	ounces
Weight of 30-round magazine	8	ounces
Number of grooves in barrel	6	
Rifling, R. H. one turn in	16	inches

6. CAUTIONS.

a. All rust preventive compound or other preservative must be removed from the gun before firing. To do this it is necessary to disassemble the gun completely (par. 18 and 19).

b. Before removing the frame, see that the bolt is in the most forward position, the safety at FIRE, and the rocker pivot at FULL AUTO. This is to avoid possible damage to the bolt.

c. The Thompson Submachine Gun, Caliber .45, M1 is not considered safe when a loaded magazine is in place and the bolt is forward in the receiver, unless the manual safety provided is turned to the "SAFE" position. When the gun is dropped on the butt from a height of 18 inches or more, the inertia of the bolt is sufficient to move the bolt rearward far enough to pick up a round and fire it.

Section III

OPERATION AND FUNCTIONING

7. LOADING THE 20-ROUND MAGAZINE.

Press down a cartridge on the magazine follower until it is retained by the mouth of the magazine. Repeat the operation until the magazine is loaded. The cartridges should feed easily into the magazine.

8. LOADING THE 30-ROUND MAGAZINE.

Proceed as in paragraph 7 above.

9. LOADING THE SUBMACHINE GUN.

Retract the bolt handle to cock the weapon. Turn the safety to SAFE. Push the loaded magazine up into the groove of the trigger frame (fig. 3), until the magazine catch can snap into position and hold the magazine securely.

10. FIRING THE SUBMACHINE GUN.

a. **Single Fire (Semiautomatic).** Turn the rocker pivot to SINGLE and the safety to FIRE. Squeeze the trigger to fire one round. It is necessary to squeeze the trigger each time a single round is to be fired. Therefore, release the trigger quickly after each shot.

b. **Full Automatic Fire.** Turn the rocker pivot to FULL AUTO and the safety to FIRE. Squeeze the trigger to fire a burst. The gun will continue to fire automatically as long as the trigger is held and there is ammunition in the magazine.

11. UNLOADING THE SUBMACHINE GUN.

When the magazine has been emptied, the bolt is automatically held in the open position. To close the bolt on an empty chamber proceed as follows. First rotate the magazine catch counterclockwise and remove the magazine (fig. 4). Then grasp the bolt handle in re-

OPERATION AND FUNCTIONING

Figure 3 — Attaching the Magazine

THOMPSON SUBMACHINE GUN, CAL. .45, M1

MAGAZINE CATCH RETRACTED

RA PD 50290

Figure 4 — Removing the Magazine

tracted position and squeeze the trigger, allowing the bolt to go slowly forward on an empty chamber.

12. ACTION OF TRIGGER MECHANISM.

a. Rocker Pivot Set at SINGLE. When the trigger is squeezed, the trigger rotates around the trigger pivot (the forward pin of the pivot plate) and lifts the disconnector up under the sear lever. The sear lever lifts the front end of the sear. This causes the sear to rotate around the sear pivot (the rear pin of the pivot plate), and in so doing depresses the nose of the sear, disengaging it from the sear notch on the under side of the bolt. As the bolt goes forward, the point of the rocker is in the T-groove on the under side of the bolt. When the point of the rocker strikes the rear end of the T-groove, the rocker is forced forward. The rounded part of the rocker comes in contact with the disconnector and forces the disconnector out from under the sear lever. As soon as the disconnector has been disengaged from the sear lever, the sear spring forces the nose of the sear up so that the sear notch on the bolt will catch on the next forward movement of the bolt.

b. Rocker Pivot Set at FULL AUTO. The rocker pivot is of eccentric design, so that when the rocker pivot is set at **FULL AUTO**, the rocker is lowered enough to allow the bolt to move forward without striking the point of the rocker. Therefore, the disconnector will not be

8

disengaged from the sear lever and the sear will remain in its lowered position as long as the trigger is depressed.

c. **Safety Set at FIRE.** When the safety is turned toward the front, the flat milled surface is in such a position that the sear is allowed to rotate around the sear pivot, thus depressing the nose of the sear.

d. **Safety Set at SAFE.** When the safety is turned toward the rear, the rounded part of the safety engages in a groove in the rear of the sear and locks the sear in its uppermost position. The safety can be turned only when the bolt is to the rear.

e. **Trip.** As the magazine empties, a fin on the back of the magazine follower rises up under the trip, causing the trip to rotate around the trigger pivot pin, compressing the disconnector spring so that the disconnector clears the sear lever and the nose of the sear rises. Thus the bolt will not go forward on an empty chamber when the magazine is emptied.

13. BACKWARD MOVEMENT OF RECOILING PARTS.

As the cartridge is fired, the pressure of the powder gases acts against the empty cartridge case and forces the bolt backward against the action of the recoil spring. The empty case is held by the extractor against the face of the bolt. After the bolt has traveled to the rear about 2 inches, the ejector, which protrudes in a groove on the left side of the bolt, comes in contact with the base of the empty cartridge and throws it to the right through the ejector opening. The bolt still has about 1¾ inches to go to the rear before the back of the bolt comes in contact with the buffer. During the rearward movement, the bolt compresses the recoil spring and expends nearly all the energy imparted by the chamber pressure, so that the bolt does not strike heavily against the buffer. The buffer absorbs the remaining shock. On the under side of the bolt there are two sear notches, so that if the bolt strikes the buffer, the rear sear notch will pass over the sear and allow the sear to engage the front notch. If the movement is not strong enough to cause the bolt to strike the buffer, the sear will engage the rear notch. If the bolt moves to the rear far enough to eject the empty cartridge case and to feed the next cartridge from the top of the magazine, the bolt will normally be back far enough to engage the sear with the rear notch.

14. FORWARD MOVEMENT OF RECOILING PARTS.

When the trigger is squeezed, the bolt moves forward under the action of the recoil spring. After the bolt moves forward about 1 inch, the forward end of the bolt comes in contact with the back of a cartridge and pushes it forward until the nose of the bullet comes in contact with the bullet ramp in front of the receiver. The lips of the magazine hold the cartridge in a straight line until the cartridge has almost cleared the magazine. The cartridge is guided into the chamber by the bullet ramp and the lips of the magazine. When the cartridge has been seated in the chamber, the extractor snaps around the rim of the cartridge. Just before the bolt reaches its forward position, the hammer on the under side of the bolt strikes the receiver. The hammer being of a triangular shape, the lower point strikes the receiver, causing the hammer to pivot around the hammer pin and strike the head of the firing pin with the upper point, thereby firing the cartridge. The rectangular surface of the bolt, striking the abutment of the receiver, stops the forward movement.

Section IV

IMMEDIATE ACTION AND MALFUNCTIONS

15. GENERAL.

This section is intended to provide necessary instruction in the related subjects of immediate action, and malfunctions and corrections. These instructions should be studied before any firing is done by the individual.

16. IMMEDIATE ACTION.

a. Immediate action is the immediate and automatic application of a remedy to get the gun working if it jams or otherwise malfunctions while firing in actual or simulated combat. When a stoppage occurs during firing, perform the immediate action described below or such portions thereof as are required to remedy the stoppage.

b. Failure of Submachine Gun to Fire. If the loaded gun fails to fire when the trigger is squeezed, proceed immediately as follows:

(1) Wait five seconds before opening chamber.

(2) Retract or cock the bolt by a sharp, quick pull on the bolt handle to eject the round.

(3) Inspect the chamber to see that it is clear. If the chamber is clear, squeeze the trigger and fire.

(4) If the chamber contains an unexpended round, set the gun at SAFE, and remove the magazine.

(5) Turn the gun over on its side and shake it to allow the round to fall out. If the round does not fall out, remove it by pushing the cleaning rod through the bore from the muzzle end, making certain that the gun points in a safe direction. Attach the magazine and resume firing.

17. MALFUNCTIONS AND CORRECTIONS.

a. Proper care of the submachine gun before, during, and after firing will usually eliminate most stoppages. Stoppages or other malfunctions which cannot be remedied by the application of immediate action should be dealt with in accordance with instructions described in the following paragraphs.

THOMPSON SUBMACHINE GUN, CAL. .45, M1

b. Failure to Fire.

(1) CAUSES. Failure to fire is generally caused by:

(*a*) Defective ammunition.

(*b*) Defective firing pin.

(*c*) Bolt not fully closed.

(2) REMEDIES.

(*a*) If the primer of round is deeply indented, the round is defective and must be discarded.

(*b*) If the primer is not indented or only slightly indented, the firing pin may be worn or broken, or the bolt may not have been fully home. Check for dirt or any other obstruction on the bolt and in receiver which prevents the bolt from closing fully. Check for a ruptured case in the chamber. Remove the obstruction. If the recoil spring is too weak to drive the bolt fully home, it should be replaced.

(*c*) Replace the firing pin if it is worn or broken.

c. Failure to Feed.

(1) CAUSES. Failure to feed may be caused by:

(*a*) Defective magazine.

(*b*) Insufficient recoil of bolt to pick up a new round.

(2) REMEDIES.

(*a*) If the magazine does not feed cartridges into the gun because of defective spring, follower, or lips, it should be replaced.

(*b*) Insufficient recoil may be due to obstruction in the receiver. Remove the frame and eliminate the obstruction.

d. Failure to Extract.

(1) CAUSES. Failure to extract is generally caused by:

(*a*) Extremely dirty chamber.

(*b*) Extremely dirty ammunition.

(*c*) Improper assembly of the gun, such as failure to position extractor properly.

(*d*) Cartridge case chambered in a hot barrel.

(*e*) Broken extractor.

(2) ACTION.

(*a*) When failure to extract occurs, the bolt may be found fully closed with a spent case in the chamber. Generally, most failures to extract can be remedied by pulling the bolt handle smartly to the rear. If this does not remove the case, use a cleaning rod.

(*b*) Sometimes the empty case will be left in the chamber, the extractor ripping through the base of the cartridge. When this occurs,

IMMEDIATE ACTION AND MALFUNCTIONS

the bolt generally will attempt to feed a fresh cartridge into the chamber. It will then be necessary to remove this round before the spent case can be removed.

(c) Where a dirty çhamber or dirty ammunition is indicated, clean the chamber and discard or clean very dirty ammunition. Faulty assembly or broken extractor will cause recurring failures to extract. Replace missing or broken parts.

e. **Gun Fires Automatic But Not Semiautomatic.** This is due to incorrect assembly of rocker. Assemble the rocker correctly (with flat side against sear lever).

THOMPSON SUBMACHINE GUN, CAL. .45, M1

Section V

DISASSEMBLY AND ASSEMBLY

18. REMOVAL OF GROUPS AND ASSEMBLIES.

a. Magazine.

(1) Point submachine gun in a safe direction. Retract the bolt. See that the chamber is clear.

(2) Rotate the magazine catch counterclockwise and slide out the magazine (fig. 4).

b. Sling. Disconnect the sling hooks from the link and swivels and remove the sling.

c. Butt Stock Group. Unscrew the two butt stock screws and remove the stock. It is not necessary to remove butt stock group to remove frame.

d. Frame Group.

(1) Grasp the bolt handle in the retracted position with one hand and with the other squeeze the trigger, allowing the bolt to move slowly forward on an empty chamber.

(2) See that safety is at FIRE and rocker pivot at FULL AUTO.

(3) Place the gun upside down on knee or on table, barrel extending rearward.

(4) With thumb of left hand press frame latch and with right hand tap frame, sliding it rearward a short distance (fig. 5).

(5) Take gun from knee or table, grasping receiver with left hand, squeeze the trigger and slide the frame off to the rear (fig. 6).

CAUTION: If safety is not set at FIRE and the rocker pivot at FULL AUTO, the bolt may be damaged.

e. Bolt and Recoil Spring Group.

(1) Support muzzle of barrel on knee or table with open side of receiver facing the operator.

(2) Press the buffer pivot slightly into the receiver and withdraw the buffer (fig. 7). Do not release the pressure on the pilot. Gradually pull out the pilot with recoil spring (fig. 8).

DISASSEMBLY AND ASSEMBLY

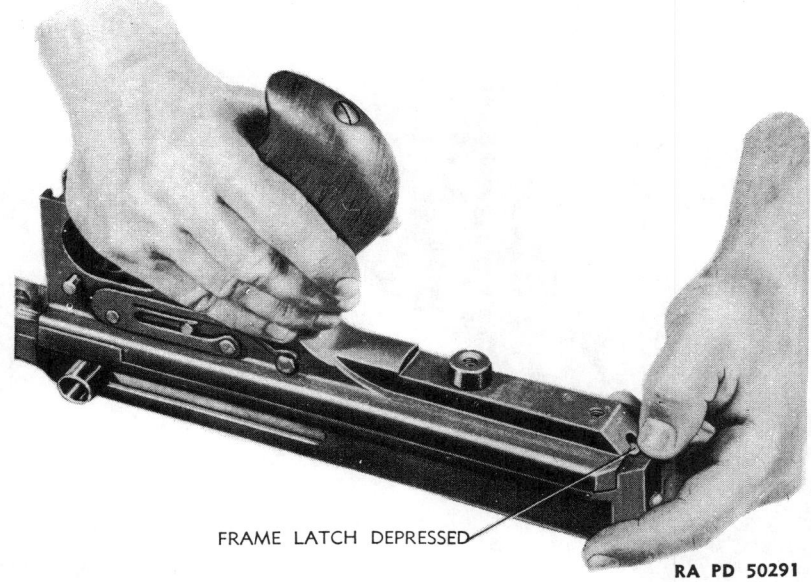

FRAME LATCH DEPRESSED

RA PD 50291

Figure 5 — Depressing Frame Latch and Tapping Frame

RA PD 50292

Figure 6 — Removing the Frame

(3) Slide the bolt rearward and tip the rear end until the bolt handle rests in the semicircular cut on the right side of the receiver. Steady the bolt in this position. Press the bottom of the hammer rearward to disengage the bolt handle and remove the handle (fig. 9).

f. Disassembled view of gun shown in figure 10.

THOMPSON SUBMACHINE GUN, CAL. .45, M1

RA PD 50293

Figure 7 — Removing the Buffer Assembly

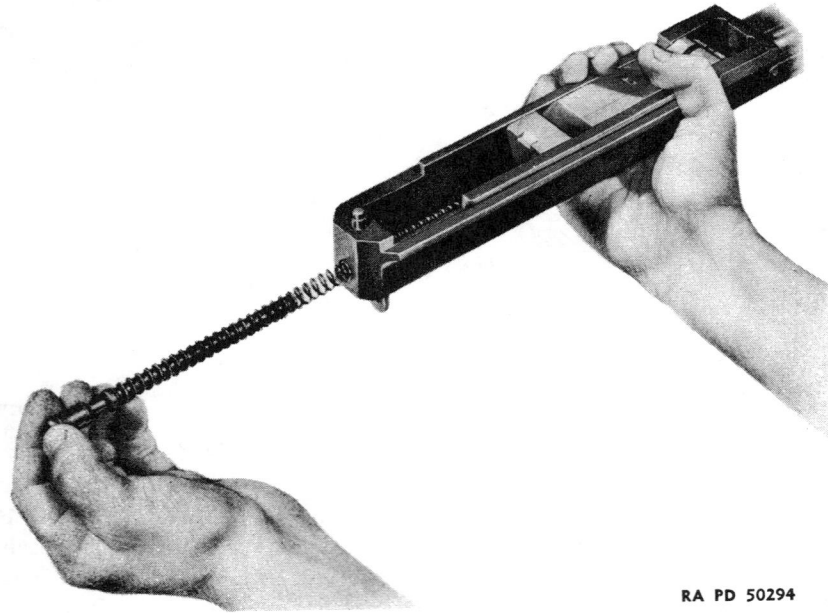

RA PD 50294

Figure 8 — Removing Recoil Spring and Buffer Pilot

DISASSEMBLY AND ASSEMBLY

RA PD 50295

Figure 9 — Removing the Bolt Handle

19. DISASSEMBLY OF GROUPS AND ASSEMBLIES.

a. Magazine. Slide out the floor plate, holding fingers over bottom of magazine tube to keep magazine spring from flying out. Withdraw the spring and the follower (fig. 11).

b. Butt Stock Group. Complete disassembly of butt stock is not necessary for ordinary cleaning. However, when necessary for replacement of broken or worn parts, butt stock plate may be removed by removal of screws holding it in place (fig. 12).

c. Barrel and Receiver Group.

(1) For ordinary cleaning the receiver and parts assembled thereto need not be disassembled. Barrel should be removed only for purposes of replacement and then only by authorized ordnance personnel.

(2) Ejector can be removed by lifting leaf sufficiently to disengage detent and unscrewing same from receiver (fig. 13). Do not try to unscrew ejector with bolt assembled and in forward position.

(3) Fore grip can be removed by unscrewing fore grip screw (fig. 14).

d. Bolt Group.

(1) Withdraw the bolt from the receiver and drift out the hammer pin. The hammer, firing pin, and firing pin spring will then tend to spring out under impulse of firing pin spring. Caution should be exercised to prevent these parts from springing away and becoming lost. Firing pin spring must not be stretched.

(2) Extractor should not be removed for ordinary cleaning or disassembling. To do so submits it to unnecessary strain and is apt to cause it to break or become set.

THOMPSON SUBMACHINE GUN, CAL. .45, M1

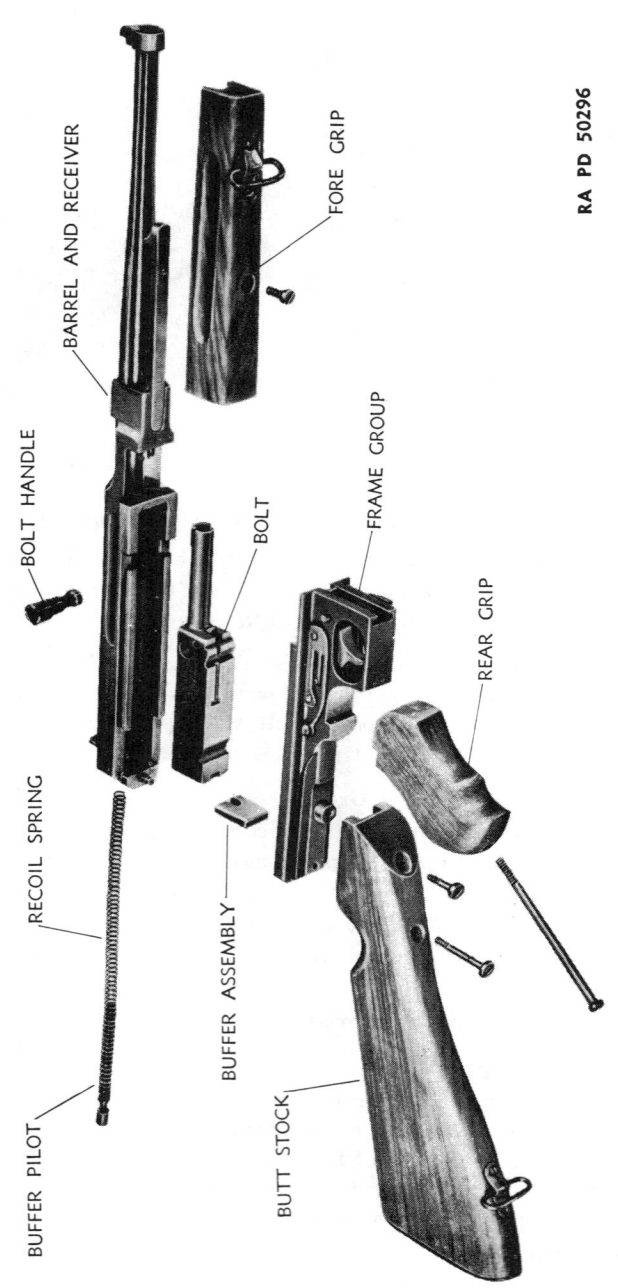

BARREL AND RECEIVER

FORE GRIP

RA PD 50296

BOLT HANDLE

BOLT

FRAME GROUP

REAR GRIP

RECOIL SPRING

BUFFER ASSEMBLY

BUFFER PILOT

BUTT STOCK

Figure 10 — Disassembled View of Gun

DISASSEMBLY AND ASSEMBLY

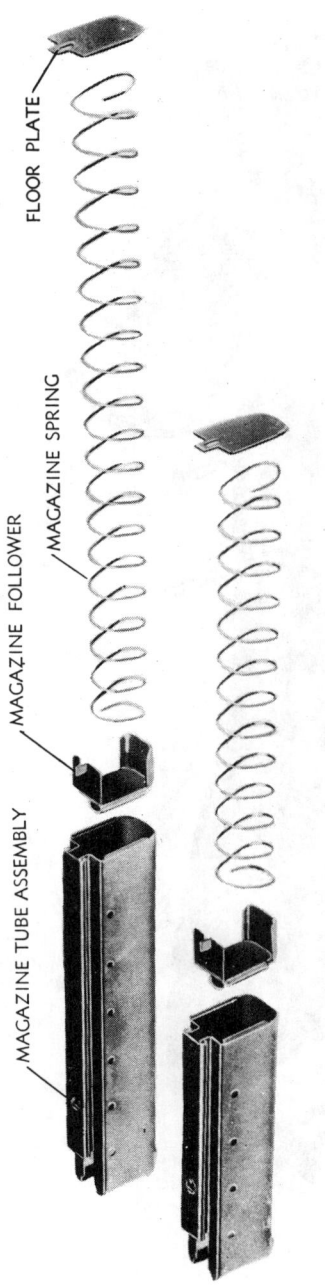

FLOOR PLATE

MAGAZINE SPRING

MAGAZINE FOLLOWER

MAGAZINE TUBE ASSEMBLY

RA PD 50297

Figure 11 — Box Magazine; 20-Round and 30-Round

THOMPSON SUBMACHINE GUN, CAL. .45, M1

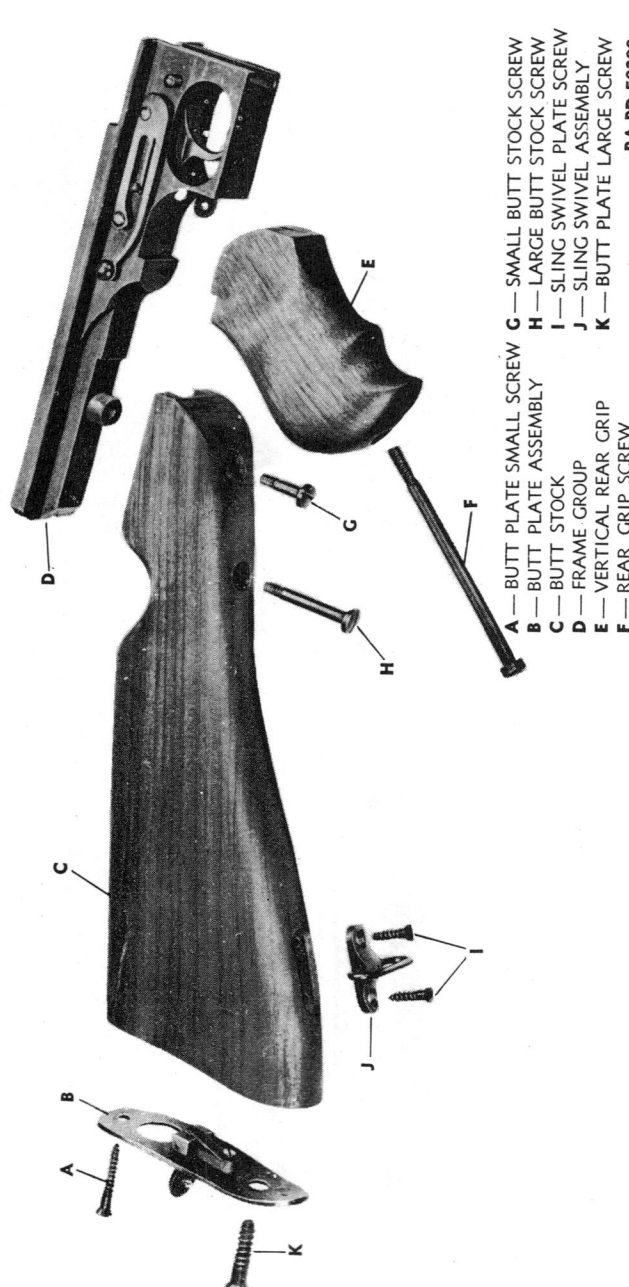

A — BUTT PLATE SMALL SCREW
B — BUTT PLATE ASSEMBLY
C — BUTT STOCK
D — FRAME GROUP
E — VERTICAL REAR GRIP
F — REAR GRIP SCREW
G — SMALL BUTT STOCK SCREW
H — LARGE BUTT STOCK SCREW
I — SLING SWIVEL PLATE SCREW
J — SLING SWIVEL ASSEMBLY
K — BUTT PLATE LARGE SCREW

RA PD 50298

Figure 12 — Frame and Stock Groups

DISASSEMBLY AND ASSEMBLY

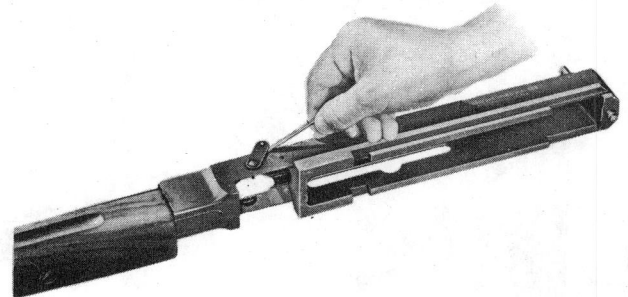

RA PD 50299

Figure 13 — Removing the Ejector

(3) To remove extractor from bolt, insert screwdriver under head of extractor on face of bolt and pull extractor out and up to withdraw it from its groove. When disassembling extractor from or assembling it to bolt, do not lift extractor higher than necessary for lug to clear anchorage hole as otherwise setting or breakage may occur (fig. 15).

e. **Frame Group** (figs. 16, 17 and 18).

(1) Hold frame in left hand, and using a screwdriver as a tool in right hand, depress short finger of pivot plate and push out rocker pivot with thumb of left hand. Lift out rocker and pivot.

(2) Again using the screwdriver but steadying hand with thumb against frame to prevent excessive movement, depress long finger of pivot plate and withdraw safety.

(3) Hold frame upright with grip in right hand. Press simultaneously with both thumbs on sear and triger pivots. These pivots project sufficiently so that by a quick pressure thereon the pivot plate will protrude on the other side far enough to permit withdrawal. While withdrawing pivot plate with left hand, press down on trigger and sear with thumb of right hand to release pressure of springs on pivots. Do not cant pivot plate during withdrawal. The remaining components of the firing mechanism are then free to be removed. Disconnector can be removed from trigger by simply withdrawing it.

(4) To remove magazine catch, rotate it in a counterclockwise direction to its full limit and lift it. Removal should be limited to replacement of broken parts. Removal of magazine catch submits magazine catch spring to unnecessary strain and is apt to damage it.

20. ASSEMBLY AND REPLACEMENT.

a. Prior to assembly, all parts must be free of dirt, rust, and other extraneous matter. Metal parts in contact must be covered with a light film of lubricating oil. In general, assembly and replacement are

THOMPSON SUBMACHINE GUN, CAL. .45, M1

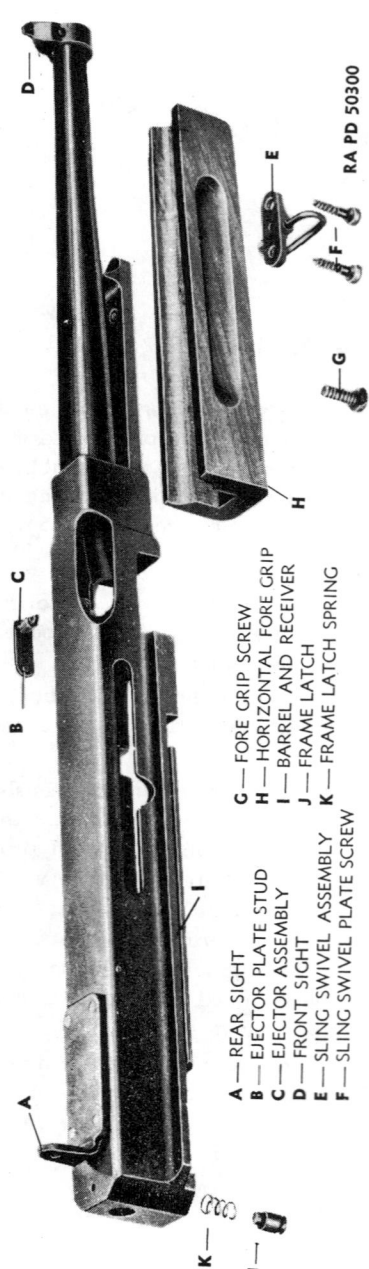

RA PD 50300

A — REAR SIGHT
B — EJECTOR PLATE STUD
C — EJECTOR ASSEMBLY
D — FRONT SIGHT
E — SLING SWIVEL ASSEMBLY
F — SLING SWIVEL PLATE SCREW
G — FORE GRIP SCREW
H — HORIZONTAL FORE GRIP
I — BARREL AND RECEIVER
J — FRAME LATCH
K — FRAME LATCH SPRING

Figure 14 — Barrel and Receiver Group

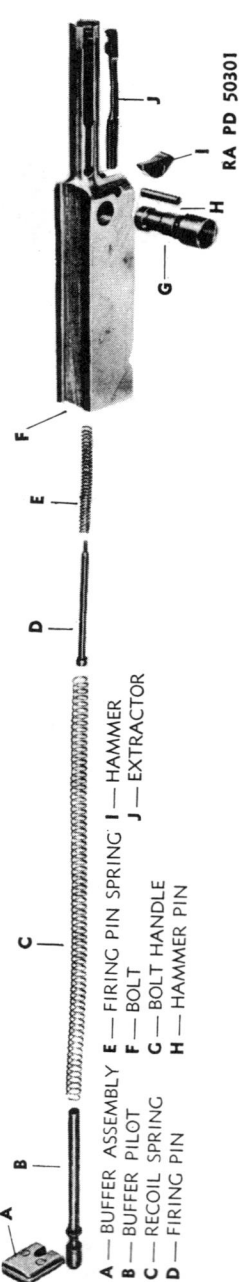

RA PD 50301

A — BUFFER ASSEMBLY
B — BUFFER PILOT
C — RECOIL SPRING
D — FIRING PIN
E — FIRING PIN SPRING
F — BOLT
G — BOLT HANDLE
H — HAMMER PIN
I — HAMMER
J — EXTRACTOR

Figure 15 — Bolt and Spring Group

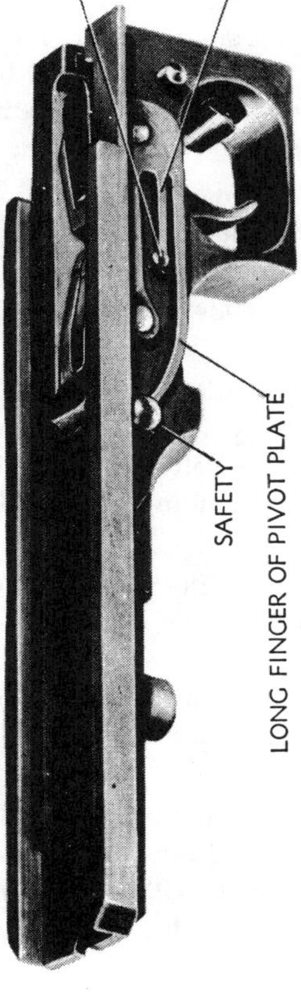

DISASSEMBLY AND ASSEMBLY

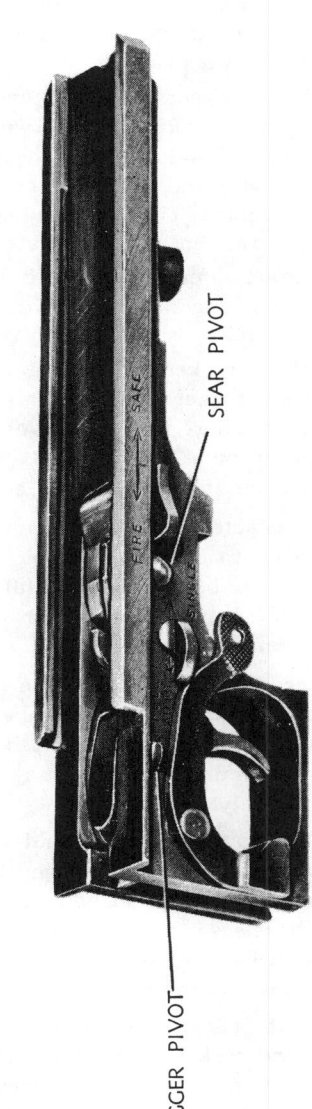

RA PD 50302

ROCKER PIVOT

SHORT FINGER OF PIVOT PLATE

SAFETY

LONG FINGER OF PIVOT PLATE

Figure 16 — Right Side View of Frame Group

RA PD 50303

SEAR PIVOT

SAFE

FIRE

TRIGGER PIVOT

Figure 17 — Left Side View of Frame Group

in reverse order of disassembly and removal. However, the following precautions should be observed in order that the parts will function properly after the gun is assembled.

b. In assembling trigger mechanism, first see that the magazine catch is in place. Assemble springs in their proper recesses. Assemble disconnector to trigger by depressing disconnector spring and sliding disconnector into place.

(1) Place trigger, trip, sear, and sear lever in their respective positions in frame, making sure forward end of sear lever rests on tip of disconnector. To aline these parts, hold frame in left hand and press downward with end of thumb on trigger and base of thumb on sear. Insert pivot plate. To avoid binding, apply gentle pressure with ball of right hand over entire pivot plate.

(2) Insert safety as far as it will go, and using screwdriver as a tool, depress long finger of pivot plate and push safety home. Turn safety to FIRE position.

(3) Place rocker in position in frame with flat side against sear lever. Insert rocker pivot as far as it will go. Using screwdriver as a tool, depress short finger of pivot plate and push rocker pivot home. Turn rocker pivot to FULL AUTO position. If rocker is assembled backward, the gun will fire full automatic but not semiautomatic.

c. If extractor has been removed, slide it into place, lifting head only enough to clear stud; avoid excessive pressure. Insert firing pin and spring in bolt, being careful to avoid stretching the firing pin spring. Place hammer in position with rounded edge upward and push hammer pin into place.

d. Place bolt in receiver, slide it forward and tilt the rear until the semicircular cut on the right side of the receiver is alined with the hole for the bolt handle on the bolt. Steady the bolt in this position. Rotate the hammer and insert the bolt handle, making sure that the handle is firmly secured.

e. Slide bolt to the rear and hold it firmly with one hand. With other hand start pilot and spring through hole in rear of receiver and push them into recess in bolt until the pilot enters the recess in bolt. Gradually release the bolt and at the same time push the pilot forward until the groove on the pilot is inside the receiver. Replace the buffer in the groove on the pilot.

f. Before fitting frame to receiver, be sure that safety is set at FIRE position and rocker pivot at FULL AUTO. Slide frame onto receiver and at the same time squeeze trigger. Frame latch will lock frame in position. Holding trigger depressed, operate bolt handle back and forth several times to test mechanism.

DISASSEMBLY AND ASSEMBLY

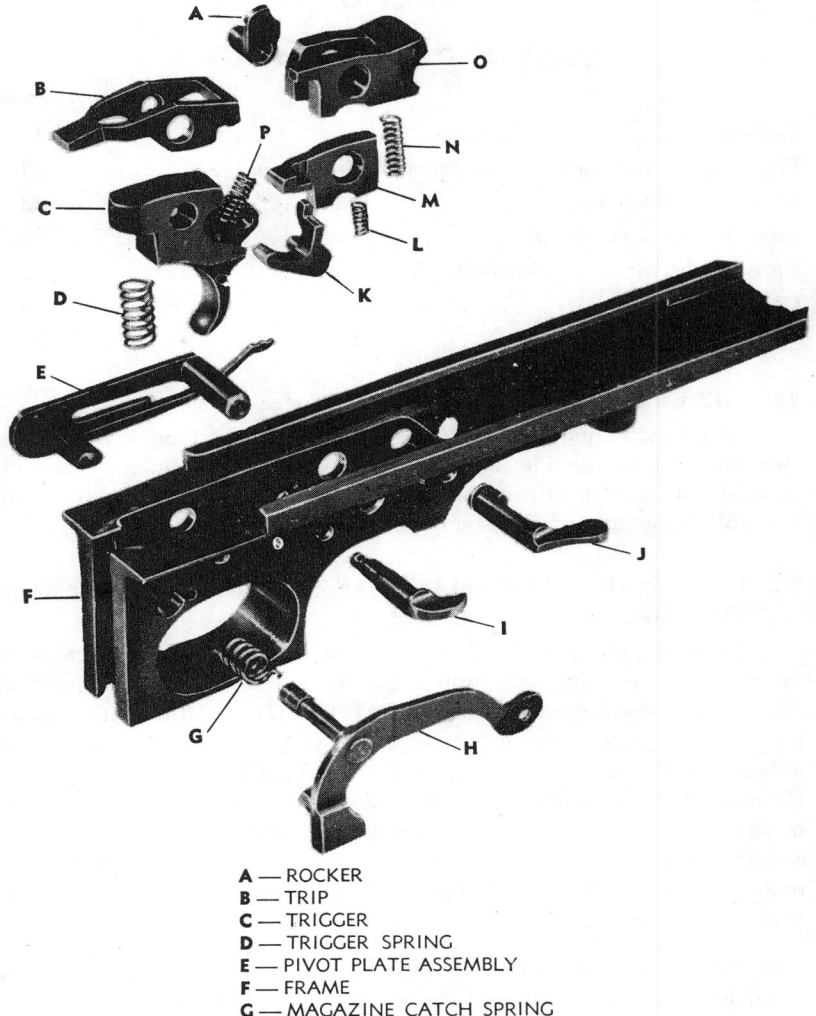

A — ROCKER
B — TRIP
C — TRIGGER
D — TRIGGER SPRING
E — PIVOT PLATE ASSEMBLY
F — FRAME
G — MAGAZINE CATCH SPRING
H — MAGAZINE CATCH
I — ROCKER PIVOT
J — SAFETY
K — DISCONNECTOR
L — SEAR LEVER SPRING
M — SEAR LEVER
N — SEAR SPRING
O — SEAR
P — DISCONNECTOR SPRING   RA PD 50304

Figure 18 — Frame Group

THOMPSON SUBMACHINE GUN, CAL. .45, M1

Section VI

CARE AND PRESERVATION

21. GENERAL.

Proper functioning and accuracy of firing depend largely on care, cleaning, and oiling. The weapon should be checked daily for cleanliness and lubrication in garrison or camp, on the range, and in the field. The following instructions should be carefully observed.

22. CLEANING OF SUBMACHINE GUN RECEIVED FROM STORAGE.

a. Submachine guns which have been stored in accordance with instructions given in paragraph 27 will be coated with either OIL, lubricating, preservative, light, or COMPOUND, rust preventive, light. Submachine guns received from storage will usually be coated with a heavy rust preventive compound. Use SOLVENT, dry-cleaning, to remove all traces of the compound. Apply the solvent with rag swabs to large parts and as a bath for small parts. Take care to remove the compound from all recesses in which springs or plungers operate. After removing all traces of the compound, allow the parts to dry and then wipe with a clean dry rag.

b. Persons handling parts after such cleaning should wear gloves to avoid leaving finger marks which are usually acid and start corrosion. SOLVENT, dry-cleaning, will attack and discolor rubber gloves.

23. CARE IN GARRISON AND CAMP.

a. Care and cleaning in garrison and camp include care of the submachine gun necessary to preserve its condition and appearance during periods when no firing is being done. Submachine guns in the hands of troops should be inspected daily for proper condition and cleanliness. Training schedules should allow time for cleaning the submachine guns daily when used in training.

CARE AND PRESERVATION

b. Bore.

(1) Remove the magazine. Retract the bolt and set the gun at SAFE. See that the chamber is clear.

(2) Assemble a cloth patch to the cleaning rod and insert the rod into the bore through the muzzle end. Run the patch back and forth several times through the entire length of the bore and chamber. Repeat with several patches until the patch comes out clean.

(3) Impregnate a patch with OIL, lubricating, preservative, light. Run the patch through the bore several times.

c. Wood and Metal Surfaces. Use a small cleaning brush to clean screw heads and crevices. With a clean dry cloth remove all moisture, perspiration, and dirt from metal surfaces and then wipe with a cloth slightly oiled with OIL, lubricating, preservative, light. This protective oil film should be maintained at all times. To clean the outer wood surfaces wipe with a cloth lightly oiled with OIL, linseed, raw. Then clean with a soft dry cloth.

d. After cleaning and protecting the submachine gun as described above, place it in the gun rack. Muzzle covers, gun covers, plugs, and rack covers should not be used because they collect moisture and promote rusting. However, when the squad rooms are being swept, it is permissible to cover the gun racks in order to protect the submachine guns from dust. As soon as the rooms have been swept, the rack covers must be removed.

24. CARE PREPARATORY TO FIRING.

a. Before firing, the following instructions should be carefully observed in order to assure proper functioning of the submachine gun.

b. Disassemble the gun into its main groups (fig. 10).

c. Run clean patches through the bore and chamber to remove all dirt and oil.

d. Thoroughly clean all metal parts and lightly oil with OIL, lubricating, preservative, light.

CAUTION: Do not oil the bore and chamber before firing because dangerous pressures may develop.

e. Lubricate the following with a drop of oil from the oiler.

(1) Extractor groove on bolt.

(2) Magazine catch.

(3) Trigger pivot.

(4) Disconnector.

(5) Sear pivot.

(6) Safety.

(7) Rocker pivot.

(8) Dovetail grooves on sides of frame and receiver.

(9) Frame latch.

f. Lubrication should be applied lightly because oil has a tendency to collect dirt which may act as an abrasive on the operating parts.

g. After the submachine gun groups have been cleaned and oiled as described above, assemble the gun and wipe all outer surfaces with a lightly oiled rag.

25. CARE ON THE RANGE AND IN THE FIELD.

a. The submachine gun must be kept free of dirt and well lubricated to obtain proper efficiency during firing. The following instructions should be carefully observed:

b. Before Firing.

(1) See that the bore is free of dust, dirt, mud, or snow.

(2) See that the chamber is clean and free of oil.

(3) Test the trigger mechanism at **SAFE** and at **FIRE**.

(4) Work the bolt back and forth rapidly several times to see that it is clean, well oiled and works freely.

(5) Examine the magazines to see that they are free of dirt and properly loaded. Discard defective magazines.

c. During Firing.

(1) Clean the chamber at reasonable intervals during extended firing in order to facilitate extraction of cartridge cases and to prevent pitting and rusting. It is not necessary to disassemble the gun for this purpose.

(2) Introduce the cleaning brush through the ejector opening in the receiver and use it vigorously.

(3) In general, it should not be necessary to disassemble the gun in the field for cleaning. However, if the mechanism becomes very dirty and functions sluggishly, disassemble the gun into its groups (fig. 10) and clean as instructed in paragraph 24.

26. CARE AFTER FIRING.

a. The weapon should be cleaned after each session of firing and not later than the evening of the day on which it was fired.

b. Immediately after firing or as soon as possible run several wet patches impregnated with **CLEANER**, rifle bore, through the bore. If the CLEANER, rifle bore, is not available, use warm soapy water or warm water alone. Remove the patch from the cleaning rod and attach the cleaning brush. Run the brush through the bore several times. Make certain the brush goes all the way through before reversing the direction. Remove the brush and run several patches wet with clean water through the bore and chamber again. Follow this with dry patches until they come out clean and dry. Finally run a patch impregnated with OIL, lubricating, preservative, light, through the bore and chamber.

c. After the bore and chamber have been cleaned, disassemble the gun completely. Clean all the metal parts with a clean, dry rag, then wipe with a lightly oiled rag before assembling. After assembling, wipe the exterior surfaces of the gun with a dry cloth. Wipe all metal surfaces with OIL, lubricating, preservative, light, and the stock and hand guard with OIL, linseed, raw.

27. PREPARATION FOR STORAGE.

a. OIL, lubricating, preservative, light, is the most satisfactory oil for preserving the mechanism of submachine guns. This oil is satisfactory for preserving the polished surfaces, bore, and chamber for from two to six weeks, depending on climatic and storage conditions. Submachine guns in short term storage should be inspected every five days. If necessary, the preservative film should be renewed.

b. COMPOUND, rust preventive, light, is satisfactory for preserving polished surfaces, bore, and chamber for a period of up to one year, depending on climatic and storage conditions.

c. Thoroughly clean with SOLVENT, dry-cleaning, the bore, all parts of the mechanism, and the exterior of the weapon. Dry with rags. After drying a metal part, the bare hands should not touch it. Then coat all metal parts with either OIL, lubricating, preservative, light, or COMPOUND, rust preventive, light, depending on the probable length of storage. The bore is best coated with rust preventive compound by dipping the cleaning brush in the compound and then running it through the bore two or three times. Then see that the bolt is fully home and, handling the weapon by the stock and guard only, place it in the packing chest.

THOMPSON SUBMACHINE GUN, CAL. .45, M1

Section VII

AMMUNITION

28. GENERAL.

The information in this section pertaining to the ammunition authorized for use in the Thompson Submachine Gun, Cal. .45, M1 includes description, means of identification, care, use, and ballistic data of the cartridges.

29. NOMENCLATURE.

Standard nomenclature is used herein in all references to specific items of issue. Its use for all purposes of record is mandatory.

30. CLASSIFICATION.

Based upon use, the principal classifications of ammunition for this gun are:

Ball — for use against personnel and light materiel targets.

Dummy — for training (cartridges are inert).

31. IDENTIFICATION.

a. General. Even though the caliber .45 cartridges are not marked or stamped to indicate the type or model, each type may be identified as described in (par 31·b). In general, the only stamping on the cartridge is that of the manufacturer's initials and the year of loading which appears on the base of the cartridge case. However, the marking on all original packing containers, both boxes and cartons, clearly and

AMMUNITION

GILDING METAL

|←1.275 MAX.→|

RA PD 4018

Figure 19 — Cartridge, Ball, Cal. .45, M1911

TIN COATED

|←1.275 MAX.→|

RA PD 4020A

Figure 20 — Cartridge, Dummy, Cal. .45, M1921

fully identifies the ammunition except as to grade (see par. 31 e). In addition to the marking, colored bands painted on the ammunition boxes and printed on carton labels provide a ready means of identification as to type (see par. 31 f).

b. Types. When removed from their original packing containers, cartridges may be identified, except as to ammunition lot number and grade, by appearance as described below and illustrated in figures 19 and 20.

(1) BALL. The bullet of caliber .45 ball ammunition has a gilded metal (copper colored) jacket. Markings appear on the bullet.

(2) DUMMY. Caliber .45 dummy ammunition is identified by its tinned cartridge case. In addition, the latest models have no primers. Earlier models have an inert primer and three 1/8-inch holes are drilled in the body of the cartridge case.

c. Model. To identify a particular design, a model designation is assigned at the time an item is classified as an adopted type. The model designation becomes an essential part of the standard nomenclature of the item and one means of identification. Prior to July 1, 1925, it was the practice to use the year in which the design was adopted as the model designation, for example, M1911. The present system of model designation is to use the letter M followed by an arabic numeral, for example, M1.

d. Lot Number. When ammunition is manufactured, an ammunition lot number, which becomes an essential part of the marking, is assigned in accordance with specifications. This lot number is marked on all packing containers. It is required for all purposes of record, including grading, use, and reports on condition, functioning, and accidents in which the ammunition might be involved. No lot other than that of current grade appropriate for the weapon will be fired (see par. 31 e). Since it is impracticable to mark the ammunition lot on each individual cartridge, every effort should be made to maintain the ammunition lot number of cartridges that have been removed from their original packings. Cartridges for which the ammunition lot number has been lost are placed in grade 3 (unserviceable ammunition which will not be fired). Therefore, when cartridges are removed from their original packings, they should be marked or tagged so that the ammunition lot number may be preserved.

e. **Grade.** Current grades of all existing lots of small-arms ammunition are established by the Chief of Ordnance and are published in Ordnance Field Service Bulletin No. 3-5. Only those lots of appropriate grades will be fired. The following grades are used in grading ammunition for this weapon:

Grade 1—For use in revolvers, automatic pistols, and submachine guns.

Grade 2—For use in automatic pistols and submachine guns only. When available, this grade is issued for these weapons in lieu of grade 1.

Grade 3—Unserviceable ammunition which will not be issued or fired.

f. **Marking.**

(1) Color bands are painted on the side and ends of the packing boxes to provide a ready means of identification as to type. Similar color stripes appear on labels on cartons, except that for dummy ammunition the label itself is green. The following color bands are used:

Ball ....Red

Tracer ....Green on yellow

Dummy ....Green

32. PACKING.

Recent lots of caliber .45 ball and dummy ammunition are packed in cartons containing 50 cartridges; 18 cartons are packed in a waxed container, and 2 containers (1,800 cartridges) are packed in a wooden box. The cubic displacement of the box is 0.77 cubic feet and its weight when filled with ball cartridges is 94.5 pounds. Its over-all dimensions are $12\%_{16}$ x $10^{15}\!\!/_{32}$ x $9^{11}\!\!/_{16}$. Older lots of ammunition are packed in 20-round cartons, 100 cartons (2,000 cartridges) being packed in a sealed metal-lined box. For additional packing data, see SNL T-2.

AMMUNITION

33. CARE, HANDLING, AND PRESERVATION.

a. Small-arms ammunition, as compared with other types of ammunition, is not dangerous to handle. Care, however, must be observed to keep the packing cases from becoming broken or damaged. All broken cases must be immediately repaired and careful attention given to the transfer of all markings to the new parts of the box. In case the box contains a metal liner, it should be air-tested and sealed, provided that the equipment for this work is available.

b. Ammunition boxes should not be opened until the ammunition is required for use. Ammunition removed from its container, particularly in damp climates, may become corroded, thereby causing the ammunition to become unserviceable.

c. The ammunition should be protected from mud, sand, dirt, and water. If it gets wet or dirty, it should be wiped off at once. Verdigris or light corrosion, if it forms on cartridges, should be wiped off. However, cartridges should not be polished to make them look better or brighter.

d. The use of oil or grease on cartridges is prohibited.

e. Ammunition should not be exposed to the direct rays of the sun, or other source of heat, for any length of time. Such exposure may affect seriously its firing qualities.

f. Whenever cartridges are taken from their original packing containers, they will be tagged or otherwise marked so that the ammunition may be identified as to lot number. Such identification is necessary to prevent otherwise serviceable ammunition from being placed in grade 3, through loss of lot number.

34. PRECAUTIONS IN FIRING.

a. Ammunition which is seriously corroded should not be fired.

b. Do not fire dented cartridges, cartridges with loose bullets, or otherwise defective rounds.

c. No caliber .45 ammunition will be fired until it has been positively identified by ammunition lot number and grade as published in the latest revision or change to Ordnance Field Service Bulletin No. 3-5.

d. Before firing, the firer should be sure that the bore of the weapon is free of any foreign matter such as cleaning patches, mud, sand, snow, and the like. To fire a weapon with any obstruction in the bore may cause the gun to burst and result in injury to the firer.

THOMPSON SUBMACHINE GUN, CAL. .45, M1

e. Do not fire oiled or greased cartridges without first removing the oil or grease, nor those which have become heated due to exposure to the direct rays of the sun or other sources of high temperature. Such cartridges, if fired, may develop hazardous chamber pressures.

35. STORAGE.

a. Whenever practicable, small-arms ammunition should be stored under cover. When necessary to leave small-arms ammunition in the open, it should be raised at least six inches from the ground and covered with a double thickness of tarpaulin. Suitable trenches should be dug to prevent water from flowing under the pile.

b. In a fire, small-arms ammunition does not explode violently. There are small individual explosions of each cartridge, the case flying in one direction and the bullet in another. In case of fire, it is advisable to keep those not engaged in fighting the fire at least 200 yards from the fire and have them lie on the ground. It is unlikely that the bullets and cases will fly over 200 yards.

c. Small-arms ammunition in storage should be protected from extreme heat in order to avoid decomposition of the propellent powder. The combination of high temperature and a damp atmosphere is particularly detrimental to the stability of the powder.

d. When only a part of a box is used, the remaining ammunition in the box should be protected against unauthorized handling and use by firmly fastening the cover in place.

36. AUTHORIZED ROUNDS.

The following ammunition of appropriate grade is authorized for use in the Thompson Submachine Gun, Cal. .45, M1. It will be noted that the nomenclature completely identifies the ammunition as to type, caliber, and model.

CARTRIDGE, ball, cal. .45, M1911.

CARTRIDGE, dummy, cal. .45, M1921.

37. BALLISTIC DATA.

The average velocity and maximum range of the **CARTRIDGE,** ball, cal. .45, M1911 as used in the Thompson Submachine Gun, Cal. .45, M1, is shown below:

Average velocity, at 25.5 feet from muzzle.... 885 feet per second

Maximum range .... 1,700 yards

38. DEFECTS FOUND AFTER FIRING.

Name of defect	How to recognize	Common causes — precautions
a. Misfire.	No action on firing. Primer shows normal impression of firing pin.	Primer is defective.
	No action on firing. Primer shows light impression of firing pin.	Indicates mechanical defect in weapon as short or broken firing pin, weak firing pin spring, bolt of weapon not being fully home, or grease in firing pin hole which cushions blow of firing pin; or caused by defective cartridge or primer.
	No action on firing. Primer shows normal impression of firing pin, but off center.	Defect in weapon.
b. Hangfire.	Delayed ignition of powder in the cartridge.	Small or decomposed primer pellet, damp or light blow of firing pin caused by dirt or defect in weapon. This is a serious defect if delay is long enough to permit the bolt to be opened before the powder burns completely, in which case injury to firer or damage to weapon, or both, may result.

THOMPSON SUBMACHINE GUN, CAL. .45, M1

Name of defect	How to recognize	Common causes — precautions
c. **Pierced Primer.**	Perforation of primer cup by the firing pin. Discoloration around indent of very small perforation. Disk from large perforation blown into action of gun, with such an escape of gas as to lower velocity of the shot.	Imperfect firing pin or very thin metal in base of primer cup.
d. **Primer Leak.**	Discoloration around the primer and the head. Slight discoloration when primer leak is small, or heavy for a large primer leak.	Too small a primer, too large a primer hole, or excessive pressure generated by propelling charge.
e. **Blown Primer.**	Primer is blown completely from pocket of cartridge case.	Serious defect but seldom encountered.
f. **Primer Setback.**	Primer protrudes above the metal head.	Defective bolt or cartridge, or excessive pressure.
g. **Leak Back of Case.**	Discoloration along body of cartridge case.	Escape of gas into the action of weapon.
h. **Failure of Case to Extract.**	Failure of case to extract.	Defective extractor or cartridge.
i. **Blowback.**	Escape of gas to the rear.	Pierced primer, primer leak, blown primer, and ruptured cartridge case.
j. **Split Neck.**	Neck of case splits and is accompanied by escape of gas to the rear, upon firing.	Not to be confused with a split neck due to season cracking which can be observed before firing.
k. **Split Body.**	Longitudinal split in body of case, thereby reducing velocity of the shot.	Case of body made of defective material or has a deep draw scratch.

AMMUNITION

Name of defect	How to recognize	Common causes — precautions
l. Stretch.	Continuous ring around the body of a fired cartridge case.	Generally due to improper head space.
m. Complete Rupture.	Circumferential separation completely around body of cartridge case, causing it to separate into two parts.	Bad bolt closing, excessive head space, or defective cartridge case. This is a serious defect, because if the forward portion of case remains in the chamber after extraction, it will cause the next round to jam.
n. Partial Rupture.	Partial circumferential separation completely around body of cartridge case.	See *Complete Rupture.*

39. FIELD REPORT OF ACCIDENTS.

Any serious malfunctions of ammunition must be reported promptly to the ordnance officer under whose supervision the material is maintained or issued (see par. 7, AR 45-30).

Section VIII

ORGANIZATION SPARE PARTS AND ACCESSORIES

40. ORGANIZATION SPARE PARTS.

a. The parts of any submachine gun will in time become unserviceable through breakage or wear resulting from continuous usage, and for this reason spare parts are supplied. These are extra parts provided with the submachine gun for replacement of the parts most likely to fail, for use in making minor repairs, and in general for care of the submachine gun. They should be kept clean and lightly oiled to prevent rust. Sets of spare parts should be kept complete at all times. Whenever a spare part is taken from the set to replace a defective part in the submachine gun, the defective part removed should be repaired, or a new one procured, and replaced in the spare parts set as soon as possible. Parts that are carried complete should, at all times, be correctly assembled and ready for immediate assembly to the submachine gun. The allowance of organization spare parts for the submachine gun is prescribed in SNL A-32.

b. With the exception of replacements of spare parts mentioned in paragraph 40 a, repairs or alterations of the submachine gun by using arms are prohibited.

41. ACCESSORIES.

a. General. Accessories include the tools required for assembling, disassembling, and cleaning the submachine gun, also the gun sling, covers and similar articles. Accessories should not be used for purposes other than those for which they are intended, and when not in use should be stored in places or receptacles provided for them. There are a number of accessories, the names or general characteristics of which indicate their uses or application. Therefore, detailed descriptions or methods of use of such items are not outlined herein. Accessories of a special nature or those which have special uses are described below.

b. Chamber Cleaning Brush M6. The brush consists of a steel wire core with bristles, the core being twisted in a spiral to hold the bristles in place. It is used to clean the chamber of the submachine gun.

c. Cleaning Brush Cal. .45, M5. The brush consists of a brass wire core with bristles and tip. The core is twisted in a spiral and holds the

ORGANIZATION SPARE PARTS AND ACCESSORIES

bronze bristles in place. The brass tip which is threaded for attaching the brush to the cleaning rod is soldered to the end of the core.

d. Accessory and Spare Parts Case M1918. This is a leather box-shaped case, approximately 3¼ inches wide, 3½ inches high, and 5½ inches long. It is used to carry spare parts and a number of the smaller accessories.

e. Fabric Envelope. This is an olive-drab cotton duck envelope 3 x 3⅛ inches. It is used to carry tools and accessories.

f. 20-Round and 30-Round Magazine. This is a flat rectangular steel box designed to hold twenty (or thirty) rounds of caliber .45 ammunition. The magazine consists of the body, follower, strip, spring, and plate. The strip is fastened to one side of the magazine to provide space for the lip of the follower to slide in. The strip also is provided with a hole for the magazine catch of the gun to hold the magazine in place on the gun. The follower and springs are inserted into the magazine from the bottom, and the plate is slipped in place at the bottom of the magazine to hold follower and spring in place. The follower is provided with a lip that catches the top of the body and keeps the follower from coming out when the last cartridge has been ejected from the magazine.

g. Oiler. The oiler is a three ounce flat-circular oil can with a spout and cap. It is carried in the butt end of the stock.

h. Submachine Gun Cleaning Rod. This consists of a long steel rod having a circular loop at one end and a cleaning patch slot at the other end. Permanently affixed to the cleaning-patch end of the rod is a head having a threaded hole to receive the cleaning brush, cal. .45, M5.

i. Gun Sling M1923 (Webbing). The gun sling is fastened to the swivel provided on the gun. It consists of a long and short strap, either of which may be lengthened or shortened as desired to suit the particular soldier using it.

j. Thong. The thong consists of a tip with cleaning patch slot and a weight tied to the ends of a 30-inch length of cord. It is used in cleaning the bore of the submachine gun.

THOMPSON SUBMACHINE GUN, CAL. .45, M1

Section IX

INSPECTION

42. GENERAL.

Inspect the submachine gun at intervals for operation and functioning. In all such inspections, use dummy ammunition. The use of live ammunition is prohibited.

43. SUBMACHINE GUN AS A UNIT.

a. Check the gun for general appearance, metal parts for scratches, rust, or wear and wooden parts for cracks and nicks.

b. Note if fore and rear grips are firmly attached.

c. Retract the bolt and note any sluggish movement or binding. Remove the magazine and see that the chamber is clear. Grasp the bolt handle in the retracted position and squeeze the trigger, allowing the bolt to go slowly forward on an empty chamber. Note any binding or sluggish movement.

d. Test functioning of the magazine catch. Load the magazine with several dummy rounds and attach it to the gun.

e. Retract the bolt, se' 'he rocker pivot in either **SINGLE** or **FULL AUTO** and the safety in **SAFE**. Squeeze the trigger. The bolt should remain cocked.

f. Turn the safety to **FIRE** and squeeze the trigger. The bolt should move forward. Load the dummy cartridge in the chamber and **FIRE** it. Retract the bolt and note any difficulty or failure to extract as well as ease of ejection.

44. BARREL AND RECEIVER GROUP.

a. Remove the bolt and spring from the receiver. Hold the barrel up to the light and inspect the chamber and bore for wear, pits, or bulges. To facilitate inspection, place a piece of white paper in the receiver so as to reflect light into the bore and then rotate the barrel slowly so that the light follows the circumference of the bore. If the barrel has pits or bulges, it should be turned over to ordnance maintenance personnel for inspection and repair.

b. Check the ejector for looseness. If the ejector is loose or damaged, turn the receiver over to ordnance maintenance personnel for repair.

45. BOLT AND SPRING GROUP.

a. Examine the bolt surface for rust, roughness, or foreign matter. Inspect the sear notch, edges, corners, and grooves for burs and wear.

b. Inspect the firing pin for wear and deformation.

c. Inspect the extractor for set and deformation.

d. Check the recoil spring for kinks, fracture and lost tension.

46. FRAME GROUP.

Turn the safety to **FIRE** and the rocker pivot to **SINGLE**. Squeeze the trigger. The disconnector should lift the front of the sear lever and depress the nose of sear. Do not release pressure on trigger. Push the rocker forward. The rocker should disengage the disconnector from under the sear lever. Turn the rocker pivot to **FULL AUTO**. Squeeze the trigger and hold it. The disconnector should not be disengaged from the sear lever and nose of sear should remain depressed.

47. BOX MAGAZINE.

a. Check box magazine for fit and retention in receiver.

b. Depress the follower and note smoothness of operation and tension of spring.

c. Inspect the magazine tube for dents, cracks, deformed lips, and foreign matter. Check follower for deformation, wear and burs, and the spring for set and fracture.

Section X

MAINTENANCE UNDER UNUSUAL CONDITIONS

48. GENERAL.

When operating under unusual conditions such as tropical or arctic climates, severe dust or sand conditions, and near salt water, the precautions listed below should be scrupulously observed.

49. CARE IN ARCTIC CLIMATES.

a. In temperatures below freezing, and particularly in arctic climates, it is essential that all moving parts be kept absolutely free of moisture. It has also been found that excess oil on the working parts will solidify to such an extent as to cause sluggish operation or even complete failure.

b. The metal parts of the submachine gun should be taken apart and completely cleaned with SOLVENT, dry-cleaning, before use in temperatures below 0 F. The working surfaces of parts which show signs of wear may be lubricated by rubbing with an oiled cloth. At temperatures above 0 F, the submachine gun may be oiled lightly after cleaning by wiping with a slightly oiled cloth, using OIL, lubricating, preservative, light.

c. Immediately upon bringing indoors, the submachine gun should be thoroughly oiled, using OIL, lubricating, preservative, light, because moisture condensing on the cold metal in a warm room will cause rusting. After the submachine gun has reached room temperature, it should be wiped free of condensed water vapor and oiled again.

(1) If the submachine gun has been fired, it should be thoroughly cleaned and oiled. The bore may be swabbed out with an oiled patch and when the weapon reaches room temperature, thoroughly cleaned and oiled as prescribed in paragraph 26.

(2) Before firing, the submachine gun should be cleaned and oil removed as prescribed in paragraph **b** above. The bore and chamber should be entirely free of oil before firing.

50. CARE IN TROPICAL CLIMATES.

a. Tropical Climates.

(1) In tropical climates where temperature and humidity are high,

MAINTENANCE UNDER UNUSUAL CONDITIONS

or where salt air is present, and during rainy seasons, the submachine gun should be thoroughly inspected at frequent intervals and kept lightly oiled when not in use. The groups should be removed at regular intervals and, if necessary, disassembled sufficiently to enable the drying and oiling of parts.

(2) Care should be exercised to see that unexposed parts and surfaces are kept clean and oiled.

(3) In hot climates, OIL, lubricating, preservative, light, should be used for lubrication.

(4) Wood parts should also be inspected to see that swelling due to moisture does not bind working parts. In such cases, shave off only enough wood to relieve binding. A light coat of OIL, linseed, raw, applied at intervals and well rubbed in, with the heel of the hand, will help to keep moisture out. Allow oil to soak in for a few hours and then wipe and polish wood with dry, clean rag.

NOTE: Care should be taken that linseed oil does not get into mechanism or on metal parts as it will gum up when dry. Stock and hand guard should be removed when oil is applied.

b. Hot, Dry Climates.

(1) In hot, dry climates where sand and dust are apt to get into the mechanism and bore, the submachine gun should be wiped clean daily or more often, if necessary. Groups should be removed and disassembled as far as necessary to facilitate thorough cleaning.

(2) Oiling and lubrication should be kept to a minimum, as oil will collect dust which will act as an abrasive on the working parts and foul the bore and chamber. OIL, lubricating, preservative, light, is best for lubrication where temperatures are high, and should be lightly applied only to the surfaces or working parts showing signs of wear.

(3) In such climates, wood parts are apt to dry out and shrink, and a light application of OIL, linseed, raw, applied as in subparagraph 50 a (4), will help keep wood in condition.

(4) Perspiration from the hands is usually acid and causes rust. Metal parts should therefore be wiped dry frequently.

(5) During sand or dust storms the breech and muzzle should be kept covered.

Section XI

MATERIEL AFFECTED BY GAS

51. PROTECTIVE MEASURES.

a. Whenever materiel is in constant danger of gas attack, unpainted metal parts must be lightly coated with engine oil. Instruments are included among the items to be protected by oil from chemical clouds or chemical shells, but ammunition is excluded. Care must be taken that the oil does not touch the optical parts of instruments or leather or canvas fittings. Materiel not in use must be protected with covers as far as possible. Ammunition must be kept in sealed containers.

b. Ordinary fabrics offer practically no protection against mustard gas or lewisite. Rubber and oilcloth, for example, will be penetrated within a short time. The longer the period during which they are exposed, the greater the danger of wearing these articles. Rubber boots worn in an area contaminated with mustard gas may offer a grave danger to men who wear them several days after the bombardment. Impermeable clothing will resist penetration for more than an hour, but should not be worn for a longer period of time.

52. CLEANING.

a. All unpainted metal parts of materiel that have been exposed to any gas except mustard and lewisite must be cleaned as soon as possible with SOLVENT, dry-cleaning, or ALCOHOL, denatured, and then wiped dry. All parts should then be coated with engine oil.

b. Ammunition which has been exposed to gas must be thoroughly cleaned before it can be fired. To clean ammunition use AGENT, decontaminating, noncorrosive, or if this is not available, strong soap and cool water. After cleaning, wipe all ammunition dry with clean rags. *Do not use dry powdered AGENT, decontaminating (chloride of lime), used for decontaminating certain types of materiel on or near ammunition supplies* because flaming occurs when chloride of lime comes in contact with liquid mustard.

53. DECONTAMINATION.

a. For removal of liquid chemicals (mustard, lewisite, etc.) from materiel, the following steps should be taken:

MATERIEL AFFECTED BY GAS

(1) For all of these operations a complete suit of impermeable clothing and a service gas mask must be worn. Immediately after removal of the suit, a thorough bath with soap and water, preferably hot, must be taken. If any skin areas come in contact with mustard, if even a very small drop of mustard gets into the eye, or if the vapor of mustard is inhaled, it is imperative that complete first-aid measures be given within 20 to 30 minutes after exposure. First-aid instructions are given in **TM 9-850** and **FM 21-40**.

(2) Garments exposed to mustard must be decontaminated. Impermeable clothing that has been exposed to mustard vapor only, may be decontaminated by hanging in the open air, preferably in the sunlight, for several days. It may also be cleaned by steaming for two hours. If the impermeable clothing has been contaminated with liquid mustard, steaming for six to eight hours will be required. Various steaming devices can be improvised from materials available in the field.

b. Procedure.

(1) Commence by freeing materiel of dirt through the use of sticks, rags, etc., which must be burned or buried immediately after this operation.

(2) If the surface of the materiel is coated with grease or heavy oil, this grease or oil should be removed before decontamination is begun. **SOLVENT**, dry-cleaning, or other available solvents for oil should be used with rags attached to ends of sticks.

(3) Decontaminate the painted surfaces of the materiel with bleaching solution made by mixing one part **AGENT**, decontaminating (chloride of lime), with one part of water. All surfaces should be swabbed with this solution. Wash thoroughly with water, then dry and oil all surfaces.

(4) All unpainted metal parts and instruments exposed to mustard or lewisite must be decontaminated with **AGENT**, decontaminating, noncorrosive. Mix one part solid to fifteen parts solvent (**ACETYLENE TETRACHLORIDE**). If this is not available, use warm water and soap. Bleaching solution must not be used because of its corrosive action. Instrument lenses may be cleaned only with **PAPER**, lens, tissue, using a small amount of **ALCOHOL**, ethyl. Coat all metal surfaces lightly with engine oil.

(5) In the event **AGENT**, decontaminating (chloride of lime), is not available, materiel may be temporarily cleaned with large volumes of hot water. However, mustard lying in joints or in leather or canvas webbing is not removed by this procedure and will remain a constant course of danger until the materiel can be properly decontaminated. All

THOMPSON SUBMACHINE GUN, CAL. .45, M1

mustard washed off the materiel in this manner remains unchanged on the ground, and it is essential that the contaminated area be plainly marked with warning signs before abandonment.

(6) The cleaning or decontamination of materiel contaminated with lewisite will cause the arsenic compounds to drain into the soil and poison water supplies in the locality for either men or animals.

(7) Leather or canvas webbing that has been decontaminated should be scrubbed thoroughly with bleaching solution. If this treatment is insufficient, burn or bury such materiel.

(8) Detailed information on decontamination is contained in FM 21-40, TM 9-850, and TC ·38, 1941, Decontamination.

Section XII

REFERENCES

54. STANDARD NOMENCLATURE LISTS.

Ammunition, revolver, automatic pistol and submachine guns .... SNL T-2

Gun, submachine, cal. .45, Thompson, M1928A1 and
M1 .... SNL A-32

Cleaning, preserving and lubricating materials, recoil
fluids, special oils, and similar items of issue.... SNL K-1

Soldering, brazing and welding material, gases and
related items .... SNL K-2

Tools, maintenance, for repair of small and hand
arms, and pyrotechnic projectors.............. SNL B-20

Truck, small-arms repair, M1.................. SNL G-72

Current Standard Nomenclature Lists are as tabulated here. An up-to-date list of SNL's is maintained as the "Ordnance Publications for Supply-
Index" .... OPSI

55. TECHNICAL MANUALS.

Ammunition, general .... TM 9-1900

Small-arms ammunition .... TM 9-1990

Cleaning, preserving, lubricating and welding materials and similar items issued by the Ordnance
Department .... TM 9-850

Military chemistry and chemical agents.......... TM 3-215

Ordnance maintenance procedure: Materiel inspection and repair .... TM 9-1100

Instruction guide: Small-arms data............. TM 9-2200

56. FIELD MANUALS.

Defense against chemical attack................ FM 21-40

57. TECHNICAL CIRCULARS.

Decontamination, 1941 .... TC No. 38

58. ORDNANCE STORAGE AND SHIPMENT CHARTS.

Instructions and specifications for packaging ordnance
general supplies .... IOSSC-(a)

Instructions for marking shipments of ordnance
general supplies .... IOSSC-(b)

Ordnance storage and shipment chart, Group A.... OSSC A

59. ARMY REGULATIONS.

Range regulations for firing ammunition for training
and target practice .... AR 750-10

Qualification in arms and ammunition training
allowances .... AR 775-10

60. ORDNANCE FIELD SERVICE BULLETINS.

Small-arms ammunition .... OFSB 3-5

INDEX

INDEX

THOMPSON SUBMACHINE GUN, CAL. .45, M1

PART FIVE

AUTO-ORDNANCE CORPORATION

1936

STATEMENT

DOMESTIC sales orders are accepted only from branches of the Federal and State Governments, political subdivisions thereof and, in exceptional cases and *with the approval of the Attorney General, Washington, D. C.,* from banks and corporations having regularly organized police departments. All domestic orders must be accompanied by one of the following statements, such statements to be written on stationery bearing official letterhead of purchasing organization and to be signed by an official duly authorized to do so. Orders for Thompson Submachine Guns must be accompanied by Statement No. 1 as set forth below. Orders for spare parts, equipment and ammunition for use with guns already owned by the prospective purchaser must be accompanied by Statement No. 2 as set forth below. These statements are in conformity with the expressed wishes and at the request of the Attorney General, Washington, D. C.:

STATEMENT No. 1: In consideration of the sale by Auto-Ordnance Corporation, to the undersigned of the Thompson Submachine Guns specified in the accompanying order, it is hereby understood that such guns are for official use in the enforcement of law and order, and that they will be, and will remain, the property of this office. It is expressly agreed that they may not be resold or otherwise transferred without the permission of Auto-Ordnance Corporation, 31 Nassau St., New York, N. Y. This agreement shall enure to the benefit of said Auto-Ordnance Corporation, its successors and assigns.

STATEMENT No. 2: In consideration of the sale by Auto-Ordnance Corporation, to the undersigned, of the following parts (or equipment or ammunition), to wit:

(Here insert list of parts, etc., purchased)

it is hereby understood that such parts (or equipment or ammunition) are to be used only upon Thompson Submachine Guns in official use in the enforcement of law and order, and that such guns, whose serial numbers are _____, together with such parts (or equipment or ammunition), will be, and will remain, the property of this office. It is expressly agreed that neither the said guns nor the parts (or equipment) above referred to may be resold or otherwise transferred without the permission of Auto-Ordnance Corporation, 31 Nassau Street, New York, N. Y. This agreement shall enure to the benefit of said Auto-Ordnance Corporation, its successors and assigns.

ALL SALES MADE IN ACCORDANCE WITH THE NATIONAL FIREARMS ACT

THOMPSON ULTRA MODERN AUTOMATIC ARMS

THOMPSON GUNS are light, short, handy automatic weapons, which provide officers of the law with superior means for the protection of lives and property.

Thompson Guns are used by Police Departments, Federal Bureau of Investigation, State Constabularies, Sheriffs, Penitentiary Guards, Express and Armored Car Companies. The military models are equipment of the United States Army, United States Marines, United States Navy and the United States Coast Guard, as well as the armed forces of various foreign countries.

Being aimed from the shoulder and having a longer sight radius the Thompson is controlled better and is more accurate than a pistol fired from the extended arm. Anyone at all acquainted with firearms can easily and quickly learn to shoot the Thompson Gun with rapidity of fire, combined with great accuracy. This factor is extremely important from the Police point of view where innocent bystanders have to be considered.

Semi-automatic fire is secured by the separate pressure of the trigger for each particular shot. Semi-automatically, one hundred (100) single aimed shots can be placed in a target 30″ x 30″ square up to a range of 75 yards in sixty seconds.

In case of dire emergency the full automatic action of the Thompson Submachine Gun can be instantaneously secured by simply turning a lever. The gun will then fire like a machine gun and is easily controlled. With full automatic fire, fifty (50) shots can be readily placed in a square 48″ x 48″ up to a range of seventy-five (75) yards in ten (10) seconds time. Full automatic fire is particularly effective against fast moving or fleeting targets.

-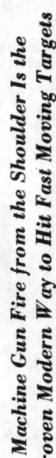

The tendency of the muzzle to rise in full automatic firing (a tendency common to all automatic weapons) is practically nullified by use of the recent invention, known as the Cutts Compensator, which reduces the recoil and virtually eliminates

*Machine Gun Fire from the Shoulder Is the
Proven Modern Way to Hit Fast Moving Targets*

THOMPSON ULTRA MODERN AUTOMATIC ARMS

all upward jump or climb of the gun when firing. This is accomplished by utilizing the gas at the muzzle in such a manner that the muzzle of the gun is returned to its original position. Though recoil in Thompson Guns is less than in high power guns, the use of the Cutts Compensator reduces the remaining recoil to practically nothing.

Semi automatically, a single Thompson Gun is capable of exceeding the combined fire effect of many older type hand loaded weapons in common use today. Full automatic fire is of still greater effect. Rapidity of fire can be regulated to meet requirements, affording the operator absolute fire control at all times.

Should the full-automatic feature of the submachine gun not be desired, the Thompson Semi-automatic Carbine can be obtained. This fires only a single shot for each separate

pull of the trigger. These separate individual shots can be fired, perfectly controlled, up to a rate of one hundred (100) shots per minute.

In a weapon used for defense of lives and property a dependable positive action is of vital importance. In both the Thompson Submachine Gun and Carbine, the Auto-Ordnance Corporation guarantees the utmost dependability. At governmental tests Thompson Guns have fired thousands upon thousands of rounds of ammunition with a remarkable freedom from misfires, jams, breakages and overheating.

It is in the field of law-enforcement that the Thompson Gun finds its most diversified usefulness. The ever mounting crime wave has increased the need for a light, fast shooting, accurate, police weapon. Both city and rural officers are frequently called upon to pursue and give battle to bandit gangs. For such

THOMPSON ULTRA MODERN AUTOMATIC ARMS

work ordinary police pistols are of little use. But a Thompson Sub-machine Gun in trained hands will more than offset the superior numbers and high powered car of the bandit gang. Not only in chasing bandits, but also in the less thrilling duties of guarding public and private property against mob violence, Thompson Guns will be found ideally adapted. Especially is this true in crowded city areas where inaccurate and high-velocity gunfire would be a menace to non-participants far from the scene of hostilities. Even in full automatic fire, the Thompson groups its shots accurately.

• Much of the usefulness of the Thompson Gun in the prison and law-enforcement field

is derived from its psychological value. Prison inmates are well aware of what comprises the arsenal equipment of their institution. The bolder convicts who would attempt escape or revolt in the face of slow firing, inaccurate hand weapons, are highly respectful of the withering barrage which can be laid down by submachine guns. Similarly a Thompson in the hands of a fearless police chief or sheriff will prove a tremendous deterrent to mobsters who might feel the urge to accomplish their purposes in defiance of the law.

• Much additional data of great interest will be found in the text dealing with the military model of the Thompson Gun on the next succeeding pages.

THE MILITARY USES OF THE THOMPSON SUBMACHINE GUN

Army and Navy Register, February 25, 1922: An exhaustive as well as successful government test of the Thompson Submachine Gun has been held at Springfield Armory. The gun passed the best test ever undergone by a portable firearm at Springfield.

THERE is a strong appreciation in military circles of the absolute necessity of increasing the fire power during an infantry attack at the moderate and short ranges, for it is at these ranges that the decision is usually gained or lost.

With the expected casualties, the heavy expenditure of ammunition at the longer ranges during the advance, the overheating of the high power rifles including the semi-automatics and automatics, and further, the necessary lifting of the covering fire of the machine guns and the artillery, the fire on the enemy is greatly diminished when the short ranges are reached, which is the very time when the success of the attack depends upon its augmentation.

At these comparatively short ranges, high velocity,

with its consequent overheating and difficult control is not a necessity. The same or greater stopping power may be secured by an increase in the bullet weight and a decrease in velocity, moreover the rate of fire and number of shots fired in a given time is not limited by the heating factor in a properly designed gun.

Combat conditions therefore call for an additional weapon to supplement and reenforce the diminishing fire of existing weapons at the critical stage of the combat. This weapon should have the following general characteristics:

1. It should have absolute freedom from stoppage due to overheating.

2. It should have a maximum effective range normally to 350—400 yds., with an emergency range to 600 yds. by use of tracer ammunition.

3. It should use an ammunition supply already established.

4. It should have full controllable semi-automatic and full automatic fire, and its full fire power should be capable of being delivered by one man.

5. It must be light in weight, and have extreme flexibility in fire control and direction.

6. Its rate of fire when fired full automatically should not be much in excess of 600 shots per minute.

7. It must function in a reliable manner, and if malfunctions occur due to ammunition or other causes, they must clear readily, without the use of tools, or long periods of inaction.

● The Thompson Submachine Gun, U. S. Navy Model 1928, meets and fulfils these conditions to the utmost possible degree. This model has been produced in accordance with the final and unanswerable test of actual "Battle Use."

This model uses the regular Cal. .45 pistol cartridge, its rate of full automatic fire is in the vicinity of 600 shots per minute, dependent to a certain extent on the power of the cartridge used. It is semi-automatic, or automatic at will.

The application of the *Cuts Compensator* has rendered it fully controllable by any one in both automatic and semi-automatic fire, with a training period measured by minutes. The intense and deadly fire power of this gun must be witnessed before it can be even partially appreciated.

The general position of the submachine gun in an Infantry organization is that of the arm for the squad leader. The squad leader while conducting and controlling his squad, does not fire at the longer ranges, and normally his entry into the fire fight at the shorter ranges brings in the power of one rifle only. When armed with the submachine gun his effective fire power, owing to mobility and flexibility, is greater than that of a machine gun, and this augmentation of the power of the firing line is at the rate of one man per squad.

When the attack is successful the gun will be of great

value in mopping up, and meeting the counter attack. Its employment at this stage of the attack, will to a certain extent, relieve the automatic and semi-automatic rifles and by affording them opportunity to cool will restore their fire rate to meet the demands of the counter attack. All automatics and semi-automatics, including the submachine gun, which does not overheat, will then be in condition for their full fire rate.

For all military purposes the submachine gun will permit one man to deliver an accurate, fully controlled, long sustained, semi-automatic fire of great stopping power within its range zone; and a controlled full automatic fire comparable to that of a machine gun, but with far greater flexibility.

This range zone must of absolute necessity appear in any attack or defense of any position, unless the attack is stopped and defeated at the longer ranges, an unusual occurrence.

In addition to the normal infantry use there are many tactical situations which occur entirely within the range zone of the submachine gun where its pres-

ence and use will be of the utmost value, particularly since machine guns and other automatics would most probably either not be present at all or be present only in limited numbers. Moreover, since the submachine gunner is gun crew and ammunition carrier, the presence of one man assures the presence of a gun in full operating condition, with ammunition supply and with no missing parts.

A number of situations involving its use are:

1. For use of troops carried to advance positions by transport planes, or other means, for the occupation of strategic points.

2. For outposts and strong points, and for combat and other patrols and isolated groups of every nature.

3. For advance, rear and flank guards.

4. For defense of convoys and pack trains.

5. For all night combat work of any nature whatever. (Its full controllable full automatic fire permits definite sectors to be absolutely swept.)

6. For occupation of cities and towns, and general street fighting.

7. For use against airplanes engaged in machine gunning and bombing the troop column. (Incendiary tracer.)

8. For ignition of woods, grass and other enemy cover (using the highly incendiary tracer bullet.)

9. For raids of all sorts, close combat in cover, and flank attacks on machine gun nests.

10. For use in tanks and armoured cars, in addition to the regular armament.

11. For general cavalry use, and in particular, when advantage is taken of their great mobility to occupy and hold advanced positions.

12. For airplanes, affording a lateral fire in the air in combat, and permitting the defense of the plane on a forced landing, thus affording time for its repair or destruction.

13. For the augmentation of the fire power of all troops who normally only carry the pistol.

14. For the armament of Engineers, signal and all special troops.

15. For every purpose where a maximum fire power, at short ranges, and even mid ranges, is required from a minimum number of men.

The submachine gun also has a definite place in the support of those arms, which are at present, owing to low or inflexible defensive fire power, practically defenseless when surprised at the moderately short ranges, such as:

16. Emplaced machine guns.

17. Infantry howitzers and 37 m/m guns.

18. Pack howitzers and other mountain artillery.

19. Field Artillery (75 m/m guns and 105 m/m Howitzers.)

Owing to the demands on the infantry, proper support for these arms is usually lacking. The use of the submachine gun, at the ranges at which these arms are endangered, will render them partcically self-supporting.

A study of the campaigns and actions of the Civil War, down to and including the World War, reveals the fact that successful attacks and defenses, with inconsiderable exceptions, always included the range

THE MILITARY USES OF THE THOMPSON SUBMACHINE GUN

zone of the submachine gun, and practically all decisions were obtained within that zone.

It further appears that more frequently than not the entire engagement from beginning to end occurred within the range zone of the submachine gun. The great value of the submachine gun is therefore readily apparent. Effective long range fire by other than machine guns and artillery is the exception rather than the rule.

The ability of the submachine gun actually to deliver a devastating fire at the ranges where hits are mostly made, its effect in gaining fire superiority, and in holding that superiority at ranges where its loss would be disastrous, cannot help but raise and maintain the morale of the attacking troops.

Its position in the fire fight and its special uses do not duplicate or infringe on the functions of the high powered arms, or any foreseen development of those arms.

It is an additional weapon, of low power, which meets and satisfies conditions that cannot be satisfied and met by high power weapons owing to their inherent limitations, and the inescapable conditions of short range combat.

•

To produce its full effect, it requires no change in organization, no special supply or training, or in fact any particular effort other than the furnishing of the gun, magazines, belt and pouch equipment for its one man unit.

1928 U. S. NAVY MODEL

THE U. S. Navy Model is especially recommended for military purposes. It is in use by the U. S. Army, the U. S. Navy, the U. S. Marines and the U. S. Coast Guard. The Marines have used U. S. Navy Models in the proportion of about one to every ten men in Nicaragua. The reports from the Marines fighting in the fields in Nicaragua have without exception paid highest tribute to both its serviceability and effectiveness for military purposes.

The outstanding characteristic of Model 1928 Submachine Gun, our U. S. Navy Model, is its slowed down action. When used in full automatic fire, Model 1928 fires at the rate of about 600 shots per minute, while a cyclic rate for Model 1921 is about 800 shots per minute.

The principles of design are identical and the general instructions for Model 1921 apply equally to the Model 1928. The slower action is accomplished by using a weighted actuator, a recoil spring of smaller diameter and a one-piece buffer, complete with pilot.

The U. S. Navy Model uses both the 20 capacity box magazines and 50 shot drum magazines. The Type L 50-cartridge capacity drum magazines which are marked "Wind to nine clicks" are the only drum magazines that should be used with Model 1928.

THOMPSON SUBMACHINE GUN MODEL Nos. 21A and 21AC

Selective Action—Single Shots or Bursts of Automatic Firing

LIST PRICES

Each

Model 21A—Thompson Submachine Gun, Standard Grade, complete with one Type XX 20-cartridge capacity box magazine $175.00

Model 21AC—Thompson Submachine Gun, Standard Grade, complete with one Type XX 20-cartridge capacity box magazine and with Cutts Compensator attached................. 200.00

SPECIFICATIONS

Calibre .45. Weight 9 lbs. 13 ounces. Length 33 inches. Length of barrel, with Compensator, 12½ inches; without Compensator, 10½ inches Equipped with Lyman sights and wind gauge. 20 and 50 cartridge capacity magazines. Ammunition calibre .45 Colt Automatic Pistol Ball Cartridges and calibre .45 Thompson-Peters Shot Cartridges (both ball and birdshot cartridges can be used in one and the same gun but require different magazines).

Rate of fire, including time required for changing magazines—Semi-automatic (a single shot for each separate pull of the trigger) up to the rate of 100 single aimed shots a minute—full automatic 300 shots a minute by instantaneously turning a small lever and pressing trigger for bursts as desired. The cyclic rate of fire is about 800 shots a minute.

Cutts Compensator (attached to muzzle in cut shown on opposite page) increases the rapidity and accuracy of semi-automatic fire, lessens the tendency of muzzle rising in full automatic firing and reduces the recoil.

THOMPSON AUTOMATIC CARBINE MODEL Nos. 27A and 27AC

LIST PRICES

MODEL 27A—Thompson Automatic Carbine, complete with one Type XX 20-cartridge capacity box *Each*
magazine **$175.00**

MODEL 27AC—Thompson Automatic Carbine with one Type XX 20-cartridge capacity box magazine
and with Cutts Compensator attached.. **200.00**

Semi-Automatic (Single Shot) Action Only

SPECIFICATIONS

Calibre .45. Weight 9 lbs. 11 ounces. Length 33 inches. Length of barrel with Cutts Compensator 12½ inches; without compensator, 10½ inches. Lyman sights and wind gauge. 20 and 50 cartridge capacity magazines. Ammunition calibre .45 Colt Automatic Pistol Ball Cartridges and calibre .45 Thompson-Peters Shot Cartridges (both ball and birdshot cartridges can be used in one and the same gun but require different magazines).

Rate of fire, including time required for changing magazines—Semi-automatic only (a single shot for each separate pull of the trigger) up to the rate of 100 single aimed shots a minute (does not shoot full automatically).

Cutts Compensator (attached to the muzzle of Carbine in cut shown on opposite page). This device stabilizes the gun when fired rapidly and reduces recoil to practically nothing. It is standard equipment on all Thompson Submachine Guns used by the United States Government.

THOMPSON SUBMACHINE GUN MODEL Nos. 28A and 28AC

Selective Action—Single Shots or Bursts of Automatic Firing

LIST PRICES

Each

$200.00

$225.00

$225.00

MODEL 28A—Thompson Submachine Gun complete with Type XX 20-cartridge capacity box magazine.

MODEL 28AC — Thompson Submachine Gun, Standard Model, (Vertical Foregrip), complete with one Type XX 20-cartridge capacity box magazine with Cutts Compensator.

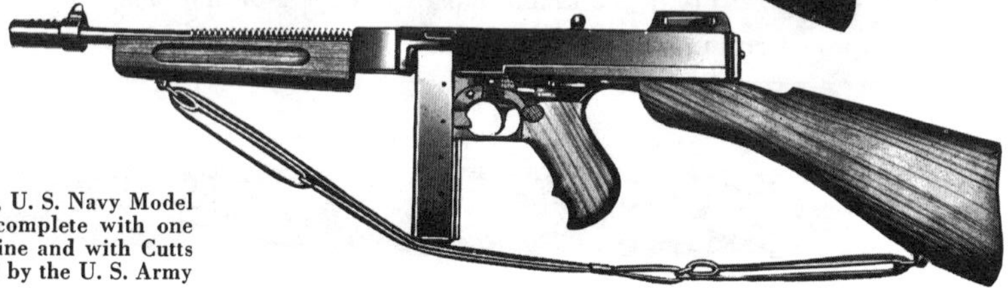

MODEL 28AC—Thompson Submachine Gun, U. S. Navy Model (Horizontal Foregrip and Sling Strap), complete with one Type XX 20-cartridge capacity box magazine and with Cutts Compensator attached. This model, is used by the U. S. Army and U. S. Navy.

SPECIFICATIONS

Calibre .45. Weight 9 lbs. 13 ounces. Length 33 inches. Length of barrel with Compensator 12½ inches; without compensator, 10½ inches. Equipped with Lyman sights and wind gauge. 20 and 50 cartridge capacity magazines. Ammunition calibre .45 Colt Automatic Pistol Ball Cartridges (230 grain bullet).

Rate of fire, including time required for changing magazines, Semi-automatic (a single shot for each separate pull of the trigger) up to the rate of 100 single aimed shots a minute—full automatic 300 shots a minute by instantaneously turning a small lever and pressing trigger for bursts as desired. The cyclic rate of fire is about 600 shots a minute.

Cutts Compensator (attached to muzzle of cut shown on opposite page) increases the rapidity and accuracy of semi-automatic fire, lessens the tendency of muzzle rising in full automatic firing and reduces the recoil to practically nothing. It is standard equipment on all Thompson Submachine Guns used by the United States Government.

THOMPSON GUN SUPER CAPACITY MAGAZINES

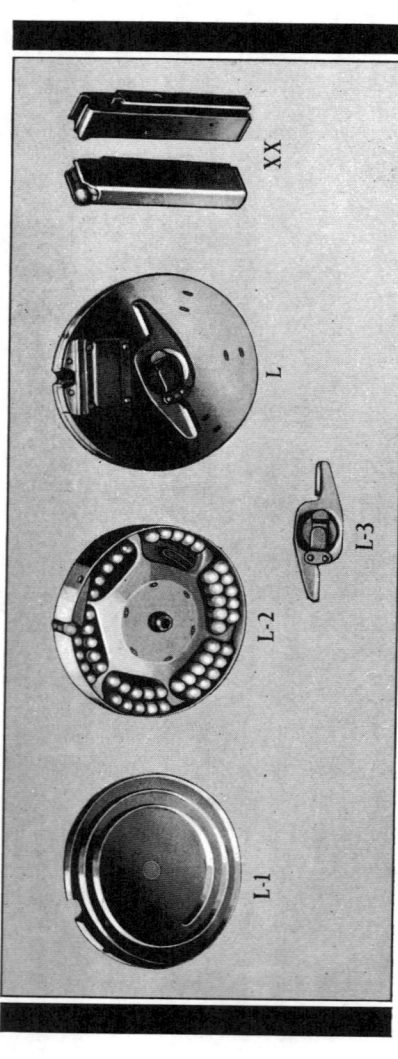

L-1   L-2   L   XX

L-3

LIST PRICES

Each

TYPE L 50-cartridge capacity drum magazine $21.00

TYPE XX 20-cartridge capacity box magazine 3.00

TYPE XVIII—Thompson Shot Magazine—18-cartridge capacity—similar in form to Type XX above—are necessary for the use of shot cartridges....................... 3.00

DESCRIPTION

L Type 50 cartridge capacity magazine assembled, L-1 Top Cover, L-2 Bottom Body Cover showing magazine loaded with 50 cartridges, L-3 Winding Key, XX Type 20-cartridge capacity magazines.

NOTE: Type XVIII 18-shot cartridge capacity magazines are almost identical in size with the Type XX Magazines illustrated on opposite page. Type XVIII Magazines are required for the bird shot cartridges used in Thompson Guns (See Ammunition Pages 22 and 23).

Magazines are interchangeable on Model 21, Model 27 and Model 28 Thompson Guns except that the Type L 50-cartridge capacity drum magazine designed specifically for Model 28 Guns is marked "Wind to nine clicks." However, the standard Type L 50-cartridge capacity drum magazine is just as well suited to the Model 28 Gun if the user will remember to wind it to nine clicks only.

CARRYING EQUIPMENT for THOMPSON GUN and MAGAZINES

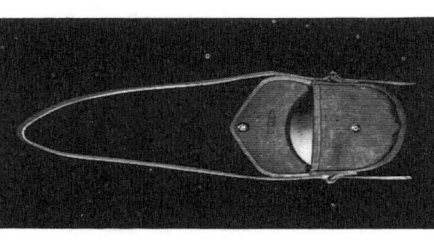

Fifty-round drum magazine case with shoulder strap.

List Price $7.00

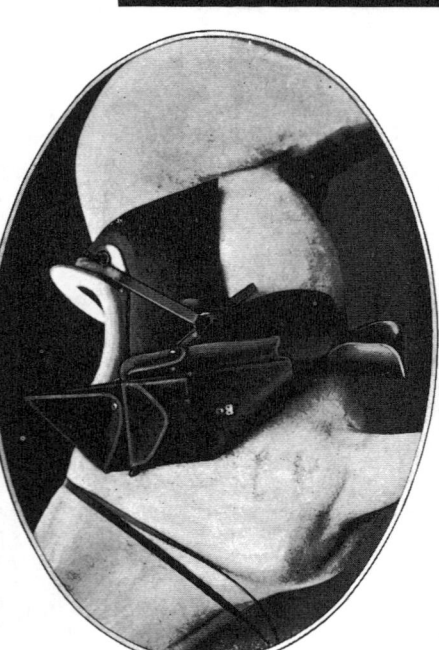

Web Gun Carrier Case for Mounted Use

List Price $16.50

Gun Carrier Case showing holster for stock and four pockets for 20-capacity box magazines.

List Price $16.50

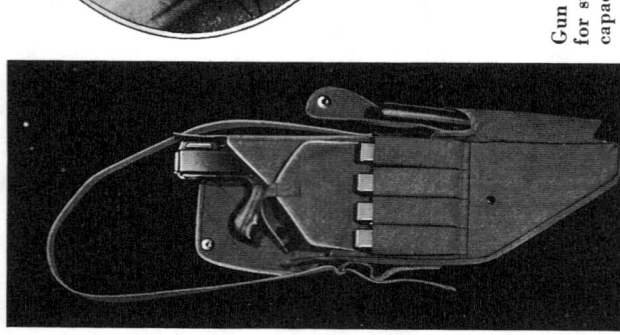

CARRYING EQUIPMENT for THOMPSON GUN and MAGAZINES

Five-pocket case for 20-round magazines and case for 50-round drum magazine attached to web belt.

List Price, Complete $13.50
Belt (separately) 2.00
50-drum Case (separately) 6.00

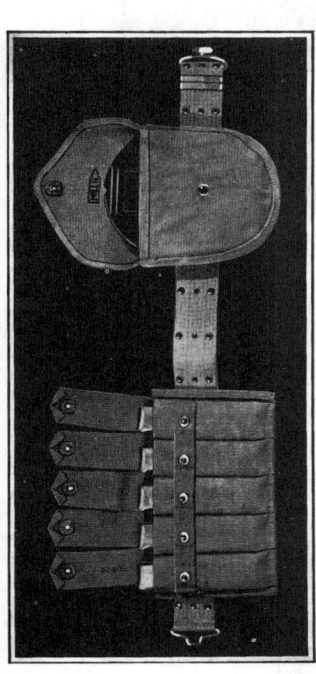

5-pocket case for 20-round magazines with separate flap on each pocket—this can also be attached to web belt...........$6.00

The gun cases and magazine containers are scientifically made up of durable waterproof material and reinforced tubing ends.

No "Getaway" Against A Thompson

No. 2
Peters-Thompson shot cartridge cal. .45
(Requires special magazine)

No. 1   *U. S. Army*
Cal. .45

AMMUNITION

Two types of ammunition are used in Thompson Submachine Guns and Thompson Carbines, namely, (see cuts on opposite page), (1) U. S. Army Calibre .45 Automatic Pistol Ball Cartridges (230 grain bullet).

(2) Peters-Thompson Calibre .45 Shot Cartridges.

The U. S. Army Calibre .45 Automatic Pistol Ball Cartridge has a velocity between 830 and 900 foot seconds; muzzle energy 430 foot pounds; penetration into white pine at ten foot range—7 inches; at 300 yards range—5½ inches; weight of cartridge 324 grains; weight of bullet 230 grains.

The 230-grain bullets used in the Thompson Gun are of such large calibre as to give them a powerful stopping effect. As pistol cartridges are of comparatively low velocity and small penetration, danger to innocent lives and property is reduced to a minimum.

The Peters Thompson Calibre .45 Shot Cartridge is somewhat longer than the regular U. S. Army Calibre .45 Automatic Pistol Ball Cartridge and requires a special type XVIII box magazine. (See pages 18 and 19, Thompson Gun Magazines.) This cartridge contains 120 No. 8 birdshot. At a range of twenty-five yards it gives an even spread of birdshot in a circle of about six feet in diameter.

Peters Thompson Birdshot Cartridges do not necessarily cause mortality, but are very useful to authorities in dealing out a lesser degree of punishment. They allow serious occasions or disorders to be handled by officers of the law in the most humane manner possible.

The Auto-Ordnance Corporation, upon request, will quote ammunition at reasonable rates on either type.

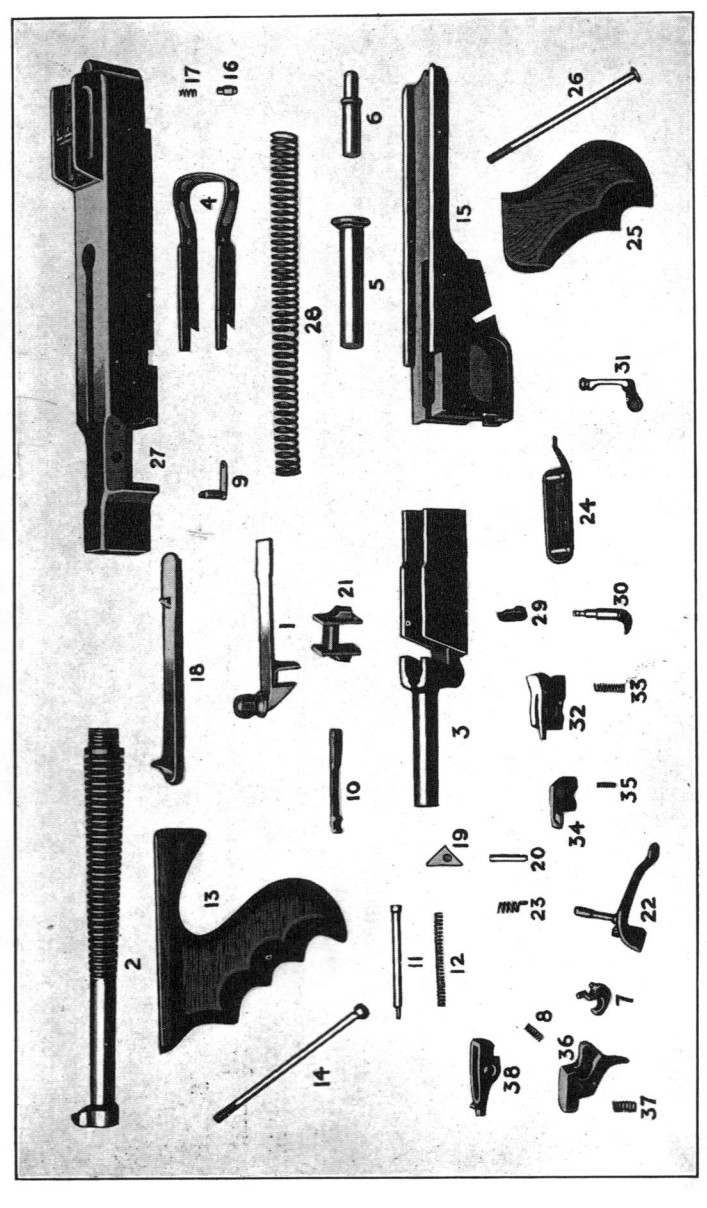

Spare parts for Model No. 27.—The spare parts for Model No. 27 are identical with those for Model No. 21 with the following exceptions:

KEY NO.	NAME OF PART	LIST PRICE
1	Actuator	$ 5.90
2	Barrel with front sight	27.50
3	Bolt	22.50
4	Breech Oiler (including felt pads)	1.75
5	Buffer (including fiber discs)	3.75
6	Buffer Pilot	1.75
7	Disconnector	2.50
8	Disconnector Spring	.50
9	Ejector	3.00
10	Extractor	3.75
11	Firing Pin	2.50
12	Firing Pin Spring	.50

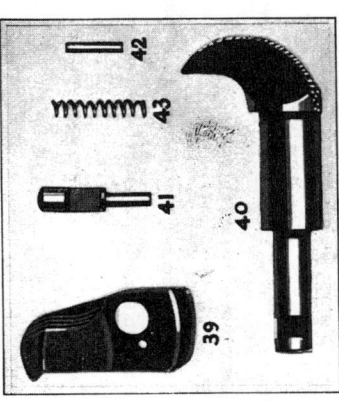

Spare parts for Model No. 28—U. S. Navy Model.—The spare parts for Model No. 28 are identical with spare parts used in Model No. 21 with the following exceptions:

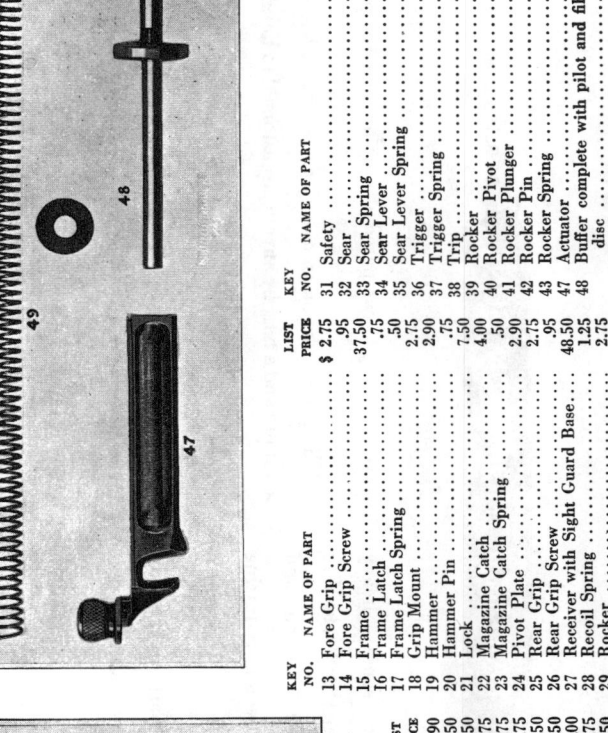

KEY NO.	NAME OF PART	LIST PRICE
13	Fore Grip	$ 2.75
14	Fore Grip Screw	.95
15	Frame	37.50
16	Frame Latch	.75
17	Frame Latch Spring	.50
18	Grip Mount	2.75
19	Hammer	2.90
20	Hammer Pin	.75
21	Lock	7.50
22	Magazine Catch	4.00
23	Magazine Catch Spring	.50
24	Pivot Plate	2.90
25	Rear Grip	2.75
26	Rear Grip Screw	.95
27	Receiver with Sight Guard Base	48.50
28	Recoil Spring	1.25
29	Rocker	2.75
30	Rocker Pivot or Fire Control Lever	2.85

KEY NO.	NAME OF PART	LIST PRICE
31	Safety	$ 2.50
32	Sear	3.50
33	Sear Spring	.50
34	Sear Lever	2.50
35	Sear Lever Spring	.50
36	Trigger	3.25
37	Trigger Spring	.50
38	Trip	2.75
39	Rocker	2.75
40	Rocker Pivot	2.85
41	Rocker Plunger	1.20
42	Rocker Pin	.50
43	Rocker Spring	.50
47	Actuator	12.90
48	Buffer complete with pilot and fiber disc	3.85
49	Recoil Spring	1.75

SPARE PARTS for THOMPSON GUNS

The Thompson Gun is not in any sense a heavy machine gun, such as was used during the World War which weighs over one hundred (100) pounds and takes a crew of six men to handle, as well as requiring many months' training.

There are only thirty-eight (38) parts to the sub-machine gun in all, and the breech mechanism is exceedingly simple, composed of only five moving parts.

The gun can be taken down and put up in less than a minute and a half by anyone accustomed to handling arms. This simplicity is greater than that of even an ordinary rifle. It requires only a few minutes shooting by a person of average intelligence to learn how to use it effectively.

The Thompson Gun spare part metal kit box container is of the same size and dimension as the Type XX 20-cartridge capacity box magazine. It will thus neatly fit in one cell of any Type XX magazine carrying case, or one of the cell pockets of the web woven gun carrier.

KEY NO.	NAME OF PART	LIST PRICE
	Spare Part Metal Kit Container	$3.50
	Brush and Thong	.50
	Short Handle Breech Cleaning Brush	.50

(Note: It is essential to keep the breech of all automatic guns in tip-top condition at all times. This brush is very handy for that purpose.)

KEY NO.	NAME OF PART	LIST PRICE
10	Extractor	3.75
11	Firing Pin	2.50
12	Firing Pin Spring	.50
20	Hammer Pin	.75
33	Sear Spring	.50
35	Sear Lever Spring	.50
37	Trigger Spring	.50

Unit price for Spare Part Kit Container, complete with above listed spare parts 13.50

BRUSH AND THONG

BREECH CLEANING BRUSH

35   33   11   37   10   20   12

CUTTS COMPENSATOR FOR THOMPSON GUN

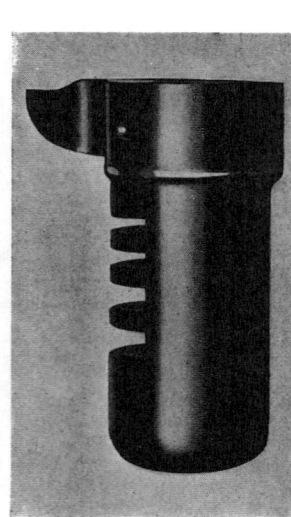

The Cutts Compensator on the Thompson Gun is so constructed that the powder gases on coming through the muzzle are coned to higher pressures which are valved through the orifices or holes in the Compensator in an upward direction, pressing the muzzle downward. This decreases the tendency of muzzle rising in full automatic firing to a very high degree. The Compensator also reduces the recoil. In semi-automatic firing, the Cutts Compensator stabilizes the Carbine, Model 27, and the Submachine Guns, Models 21 and 28, enabling greater accuracy combined with rapidity of fire. It is standard equipment on all Thompson Submachine Guns used by the United States Government. Cutts Compensators can only be fitted to Thompson Carbines and Submachine Guns at the factory.

Cutts Compensator fitted on Thompson Gun Barrel . . . . LIST PRICE $25.00

Consolidated Price List THOMPSON GUNS and ACCESSORIES

	LIST PRICE
Thompson Submachine Gun Model 21A............	$175.00
Thompson Submachine Gun Model 21AC............	200.00
Thompson Carbine Model 27A.................	175.00
Thompson Carbine Model 27AC................	200.00
Thompson Submachine Gun Model 28A...........	200.00
Thompson Submachine Gun Model 28AC— U. S. Navy Model.................	225.00
Thompson Submachine Gun Model 28AC— Standard Model....................	225.00
Cutts Compensator fitted to Thompson Gun barrel at factory—(barrel extra)................	25.00
Type XX Box Magazine—20-capacity...........	3.00
Type L Drum Magazine—50-capacity............	21.00
Type XVIII Magazine—18-capacity............	3.00
Type A—Web Gun Carrier Case for Thompson Gun without Cutts Compensator..............	16.50
Type A1—Web Gun Carrier Case for Thompson Gun with Cutts Compensator...............	16.50
Type B—Web Gun Case Carrier for Thompson Gun for mounted use without Cutts Compensator....	16.50
Type B1—Web Gun Carrier Case for Thompson Gun for mounted use with Cutts Compensator......	16.50

	LIST PRICE
Type D—Web Belt Carrier Outfit with 5-pocket Case for Type XX Box Magazine and Web Case for Type L Drum Magazine................	13.50
Belt (separately)...................	2.00
5-pocket Case (separately).............	6.00
50-drum Case (separately)............	6.00
50-drum Case for Type L Magazine with strap for carrying over shoulder............	7.00
Web Gun Sling....................	1.75
Waterproof Duck Breech Cover for Thompson Gun	4.50
Metal Kit Box Spare Part Container........	3.50
Spare Parts for Thompson Guns (see page 27)	
Gun Buttstock complete.............	17.50
Gun Sling Swivels (2) for Foregrip and Buttstock, Each......................	1.50
Brass Cleaning Wire Rod with removable wire bore brush (furnished with each gun)—extra.....	1.50
Brush and Thong for cleaning bore........	.50
Short Breech Cleaning Brush (wire handle with bristles)....................	.50
Oil Can (furnished with each gun)—extra......	.50
Hand Book of Instruction for Thompson Gun (furnished with each gun)—extra......	.50

Other Books Available From Desert Publications

PRICES SUBJECT TO CHANGE WITHOUT NOTICE

Desert Publications

800-852-4445

215 S. Washington Dept. BK- 031
El Dorado, AR 71730 USA
www.deltapress.com

Shipping & handling
1 item $4.95 - 2 or more $8.95